A First Course in ANALYTICAL MECHANICS

Klaus Rossberg

Oklahoma City University

John Wiley & Sons

New York • **Chichester** • **Brisbane** • **Toronto** • **Singapore**

Library of Congress Cataloging in Publication Data:

Rossberg, Klaus.
A first course in analytical mechanics.

Includes index.
1.   Mechanics, Analytic.   I.   Title.

QA805.R67   1983   531′.0151   83-3602
ISBN 0-471-86174-X

Printed in the United States of America

10   9   8   7   6   5   4   3   2   1

For Deanna, James and Nadya

Make the most of yourself,
for that is all there is to you.

EMERSON

preface

Many good books on analytical mechanics have been written, and most of them are still available. Therefore, the publication of another book on this subject must be well justified.

The plan for writing this book grew out of the search for a textbook that would be an introduction to analytical mechanics on the sophomore or junior level, would cover the essentials of mechanics without being too lengthy, and would have a well-balanced treatment of the foundations of mechanics and of important examples. Moreover, the book should present the necessary background material for other advanced physics courses, especially for modern physics and quantum mechanics.

Although these conditions impose strong limitations on the scope and the content of the book, they still could be satisfied in several different ways. The material selected for this book and its presentation reflect my personal views. Other physicists might include additional material and/or delete certain sections depending on their philosophies, interests, and teaching experiences.

The difficult problem of the selection of the material from the enormously large field of classical mechanics had to be attacked from both ends: Which topics are essential for the understanding of all of physics, and which areas of mechanics can be omitted from a first course in analytical mechanics without losing the broad overview? Concerning the size of the book, the second question was of greater importance. Because most sophomores have little working knowledge of partial differential equations, it was easy to decide against the inclusion of fluid dynamics, potential theory, and wave propagations. Topics that are mathematically too involved, such as the three body problem, were also omitted. Chapters on noninertial frames and Hamilton's principle were not included because these topics usually are discussed in the second semester of a sequence of mechanics courses. There remained the controversial issue of the treatment of special relativity. I decided against the inclusion of this field because a survey shows that most textbooks on introduction to physics, modern physics, and electromagnetic fields cover special relativity in approximately thirty pages. In my opinion, special relativity should be presented to physics majors in a separate one-semester course because of its importance and its broad range of applicability outside mechanics.

Returning to the first question, namely, what to include in this book, I decided to present first some background material in mathematics (Chapter 2) needed for the precise formulation of mechanics. This material is intended as a quick reference, and it should not be discussed at great length in the lecture for reasons of time. Kinematics (Chapter 3) is the logical starting point in mechanics. Dynamics (Chapter 4) is most naturally approached in Newton's formalism. Oscillations (Chapter 5) and central forces (Chapter 6) provide important examples of broad applicability, many of which are exactly solvable. Many particle systems (Chapter 7) are included because all systems consist of more than one particle in reality. Lagrange's and Hamilton's formalisms (Chapter 8) are important for the foundations of advanced mechanics and other areas of physics, for example, statistical mechanics, geometrical optics, quantum mechanics, and quantum field theory. It would be nice to introduce these formalisms at an earlier stage, but students need to have some working knowledge of Newtonian dynamics before they are exposed to more general ideas of mechanics. Although the chapter on rigid body mechanics (Chapter 9) should logically be placed right after many particle systems, some equations of rigid body mechanics are more easily derived in Lagrange's formalism, which justifies the placement of this chapter. The last chapter (Chapter 10) on coupled oscillations (which are also best treated in Lagrange's formalism) provides the basis for the transition to the theory of wave motion.

What is presented in this book is still more than what can be covered efficiently in a one-semester course. It is left to the instructor to make additional cuts.

Appendix 1 contains a list of some vector relations. The other appendixes contain material that usually is not found in elementary texts but that I considered appropriate here.

The selected problems are all of a moderate degree of difficulty, and the excessive use of mathematics has been avoided. Too much mathematics at the beginner's level may obscure the ideas of physics or frighten the students. Some questions concerning the philosophy of science and the foundations of mechanics have been included. Physics majors should not merely know that $ma = F$, they should also have some understanding of how the building of mechanics is constructed.

Analytical mechanics is logically and traditionally the first course in the sequence of theoretical physics courses because basic concepts and principles are introduced that are absolutely essential for the understanding of the other areas of physics. It is a good sign that many students are interested in the exciting frontiers of physics, but these students are advised to take the studies of the fundamentals of mechanics very seriously. Only then is it possible to fully understand the modern developments of physics that have their ultimate roots in classical mechanics. Wherever possible, I have mentioned in the text the points of departure of modern physics from classical mechanics and their relationships.

I have tried to present all material in a clear form and to avoid ambiguities, jumps, and statements such as " ... it can easily be seen that... " It is my opinion, however, that a textbook should not contain every detail of a calculation or a derivation. Spoon feeding does not help a student master his or her field of study. All physics and mathematics textbooks should not merely be read but studied with pencil and paper.

Klaus Rossberg

acknowledgments

Many people have given me their support, encouragement, and constructive criticism during the long years of preparing this book. I have learned from them in many ways, professionally and personally, and I express my gratitude to them all.

First, I thank my students who were exposed to the earlier versions of the manuscript in my mechanics courses and who continually reminded me that a textbook should be written for the students and not for the experts. I thank especially Karen Mares, Mary Tyner, and Stewart Meyer.

I thank Professors Somdev Tyagi (Drexel University), Jerry P. Draayer (Louisiana State University), Lester M. Clendenning (Humboldt State University), David B. Tribble (Loyola University of Chicago), David A. Lind (University of Colorado), Leon Blitzer (University of Arizona), and graduate student R. S. Robinson (University of Arizona), who read parts or all of the book at various stages. They made numerous helpful suggestions and also pointed out several errors. Special thanks are extended to Professors John A. Wheeler (University of Texas) and Barry N. Taylor (National Bureau of Standards) for their replies to specific questions.

It is my pleasure to express my sincerest and warmest thanks to my colleagues and friends, Professors Leonard Flansburg and Ronald Olowin, with whom I discussed details of the book almost daily, and to Professor Kurt Just (University of Arizona), who carefully examined the whole manuscript. His countless suggestions and corrections were invaluable, especially his clarifying remarks on the foundation of Newton's laws and on inertial frames. He also suggested several problems that are included in this book.

I extend my thanks to the members of the editing, designing, and production staff of John Wiley for their excellent work. I enjoyed the cordial atmosphere with the many persons I had contact with, directly or via long distance.

Finally, I thank all of my children for their great understanding of my time-consuming work and for their unselfish patience. Their love has given me the strength to finish this book.

K. R.

Oklahoma City
February 1983

contents

The more alternatives,
the more difficult the choice.

ABBE D'ALLAINVAL

10 COUPLED OSCILLATORS   243

APPENDIXES   277

SOLUTIONS TO SELECTED PROBLEMS   283

INDEX   287

chapter

1 Introduction

The begynnynge of every thynge is the hardiste.

SIXTEENTH CENTURY ENGLISH PROVERB

1.1 GENERAL CONSIDERATIONS ON PHYSICS AND MECHANICS

Since their emergence, humans have tried to explore and understand nature. Prehistoric humans sought to satisfy their basic need for survival; later, they strove to satisfy their curiosity and their philosophical needs. Part of this human endeavor is natural science.

Physics is the basis of natural science. It deals with the laws of matter over a wide range of its forms—from elementary particles and atoms to macroscopic objects, planetary systems, and the whole of our universe. It does not come as a surprise that the laws of nature are different for the many forms of matter, but it is one of the physicist's strong beliefs that a unifying principle exists which would allow him or her to derive from it the laws for all forms of matter. In spite of the progress made in recent years a complete unifying theory has not yet been found.

Physics, like all branches of natural science, is founded upon observations and experiments. A system of principles and relationships constructed to explain the observed phenomena is called a hypothesis. As a result of creative inductive reasoning, it is merely an unproven guess. A hypothesis becomes a scientific theory if it has the ability to predict new phenomena. A scientific theory is true, by definition, when it is consistently corroborated by experimental facts. It cannot be *proven* true, but it can be proven wrong. Agreement between conclusions derived from a theory and experimental confirmation can only lend support to the theory. If the conclusions disagree with an experiment, then the theory must be abandoned or amended.[1] Examples of the creation of new, enlarged theories are numerous in physics. Two of the most striking examples are the development of the theory of relativity and that of quantum mechanics.

Theoretical physics deals with the mapping of natural processes and properties of matter on a set of mathematical relations among well-defined quantities. The language of mathematics is used because it is more precise than any other

---

[1]For a short but comprehensive discussion of the philosophy of science and the foundations of physics see P. J. Brancazio, *The Nature of Physics*, Macmillan, New York, 1975.

known language and also because it allows us to express even complicated relationships in a useful shorthand notation. The mathematical frame consists of two parts, namely, the set of all definitions of the physical quantities, whether observable or not, and the set of fundamental laws and axioms from which other equations and rules can be derived. Mathematical simplicity is one of the guidelines for developing a physical theory. Simplicity is most clearly expressed in the forms of conservation principles (e.g., energy, momentum), symmetry principles[2] (e.g., the behavior of systems under certain space-time transformations), and variational principles (extremum principles). Early in this century, it became clear that the three types of principles are related to each other. The application of various symmetry principles has led to insights in elementary particle physics where dynamical laws have not yet been found.

Physics may be called an exact science in the sense that correct mathematical reasoning allows us to derive not only qualitative statements but also quantitative statements from the set of assumed fundamental laws formulated in mathematical language. This is in contrast to common sense which may be directly applied to some natural phenomena to derive qualitative statements only. But common sense may play tricks. Seemingly "obvious" assumptions can lead to conclusions that are not in agreement with the observed facts. (See Questions 1.4, 1.5, and 1.6.)

Physical quantities must be well defined, that is, they must have one and only one meaning. A definition is the reduction of a concept to be defined to other previously defined concepts. But the process of defining cannot be continued indefinitely. Some fundamental (or primitive) quantities, notions, or concepts must be accepted a priori, even in physics. For example, the concepts of space, time, and matter cannot be defined. The fundamental concepts may seem fuzzy at first, but they become clear when used repeatedly. They may also become subject to revision; for example, the notion of a flat space-time continuum turned out to be inappropriate in general relativity.

Much of our knowledge of nature is expressed in terms of the observable quantities mass, position, momentum, energy, and so on. Observable quantities are defined operationally by a measuring device, that is, by a prescription of how the quantity shall be measured. Classical physics deals directly with these observable quantities, and it is assumed that they can be measured in principle with unlimited precision. However, every attempt to apply the methods of classical mechanics to the atom failed.[3] With the work of Heisenberg, Schroedinger, and other pioneers of modern quantum mechanics it became clear that the observable quantities do not directly describe microscopic systems, and that certain combinations of observable quantities cannot simultaneously be measured with unlimited accuracy. Also, some quantities needed to be introduced that were not observables at all. Although the realm of applicability of classical physics is limited, quantities such as mass, position, momentum, and energy also have a meaning in quantum mechanics. This is because all measuring devices from which we obtain our information of nature are macroscopic systems that are subject to the laws of

---

[2]See Section 6.3.
[3]Max Born in a letter to Albert Einstein (October 21, 1921): "Die Quanten sind eine hoffnungs-lose Schweinerei." (The quanta are a hopeless mess.)

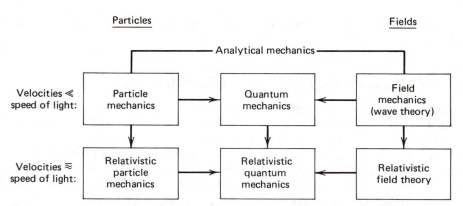

Figure 1.1 Diagram showing the relations between analytical mechanics, quantum mechanics, and relativity.

classical physics. The examples cited should alert us not to take all physical concepts for granted or to apply them uncritically to all areas of physics.

Analytical mechanics is logically and traditionally the first course in the series of theoretical physics. The name originates from Lagrange's book *Mechanique Analytique*. Following Newton's *Principia*, Lagrange derived analytically from Newton's axioms of motion the main parts of the theory of mechanics. Analytical mechanics encompasses both particle mechanics and wave theory. The relationship of analytical mechanics to other branches of physics is shown in Figure 1.1.

1.2 REFERENCE FRAMES AND COORDINATE SYSTEMS

Three important concepts need to be introduced. A physical system is a distribution of matter on which observations are made and experiments are performed. A reference frame is any part of the universe relative to which observations of a physical system are made. The existence of matter and/or radiation is necessary for a meaningful definition of the reference frame. A coordinate system is any mentally constructed system attached to a reference frame relative to which the physical system under investigation is described. Usually that particular coordinate system is chosen which yields the simplest possible description of the system and/or exhibits the largest possible degree of symmetry of the physical system. Observers in two different reference frames may or may not measure the observable quantities of a physical system with the same outcome. Even the physical laws (e.g., force laws) may not be identical, which may be the case if at least one reference frame is not inertial. (For the definition of inertial frame see Section 4.2.) However, two different coordinate systems, which are attached to the same reference frame, may still yield different descriptions of the physical system, but the physical laws must be the same in these coordinate systems because the laws of nature cannot be altered by a construction of the human mind. One often uses the term *coordinate system* very loosely. Actually, it means a "reference frame provided with a coordinate system." The reader is asked to distinguish the term *coordinate frame* from the term *reference frame* in order to avoid possible confusion.

1.3 STANDARD UNITS OF MECHANICS

All quantities in mechanics are based upon the three fundamental quantities: length, time interval, and mass. The standard international units of these quantities are the meter, second, and kilogram, respectively, defined as follows:

1 meter = 1 m = 1,650,763.73 times the wavelength of light emitted from the krypton isotope Kr^{86} in the transition from the state $2p_{10}$ to the state $5d_5$.[4]

1 second = 1 s = 9,129,631,770 times the period of oscillation between the two hyperfine levels of the ground state of the caesium isotope Cs^{133}.

1 kilogram = 1 kg = the mass of a platinum-iridium prototype cylinder which is kept at the International Bureau of Weights and Measures in Sevres, France.

Practically, it is sufficient to define the standard of length by the standard meter which is also kept in Sevres, and to define the second as 1/86 400 of the mean solar day.

1.4 ON STUDYING PHYSICS

It may be appropriate here to take a closer look at the reasons why many students experience difficulties when studying physics, especially during the first semesters. One of the difficulties is directly related to the act of learning itself. Here, learning is understood as the conscious process of acquiring knowledge or skills with the purpose that the acquired knowledge may be actively reproduced and applied when necessary. Pure memorization of equations or vocabularies is only a form of passive reproduction. A person is able to actively reproduce if the object under study has been viewed, investigated, and analyzed under several possible angles. Learning, therefore, is a process that involves time; it does not occur in an instant. All phases that a person goes through during a creative process[5] (e.g., painting a picture) from the first vague idea to the completion of his or her work also appear during the process of learning. In order to acquire certain technical skills or to understand ideas or concepts, two seemingly contradicting requirements are essential—hard work, at times even accompanied by intense feelings of frustration, followed by complete relaxation. If the facts of the learning process are well understood, the active engagement in learning becomes more enjoyable and the outcome of studying more gratifying and rewarding.

Other difficulties originate at the three steps involved in the mapping process of the phenomena and the laws of nature on the language of mathematics, and vice versa. The mapping process is outlined in Figure 1.2.

---

[4]The Comite Consultatif pour la Definition du Metre of the Comite International des Poids e Mesures is currently discussing the exact wording of a redefinition of the meter as the distance traveled by light in 1/299,792,458 of a second.

[5]See Rollo May, *The Courage to Create*, W. W. Norton & Company, Inc., New York, 1975.

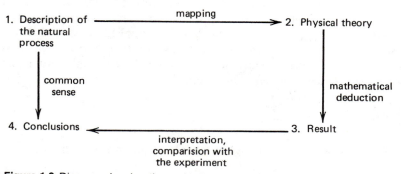

Figure 1.2 Diagram showing the path of the thought process in theoretical physics.

Step 1 to 2 is a creative process of the human mind. This is certainly true if a new theory is developed. It is also true to a lesser degree if the theory is recreated (learned), or if a set of equations needs to be found to solve an assigned homework problem. Because a creative process requires insight and intuition (which are products of the mind of the individual), a general prescription as to how to establish the mapping from nature to the physical theory (or parts thereof) cannot be given. J. A. Wheeler's "first moral principle"[6] expresses these ideas in a humorous way: "Never make a calculation until you know the answer."

Mathematical difficulties that may appear in step 2 to 3 are usually overcome as the student progresses in mathematics.

Step 3 to 4 is the comparison of the derived mathematical result with the observed data. The translation from the "foreign" mathematical language to ordinary language is usually much simpler than the inverse step 1 to 2.

The finding of the mapping 1 to 2 can be facilitated if one heeds the following suggested study hints:

1. The definitions of physical quantities should not merely be memorized mechanically. It is essential that the student understand all implications and ramifications of the defined quantity. This very important advice is usually underestimated or entirely ignored by students.

2. The student should learn to distinguish between the physical and mathematical aspects of a physical problem. The progress from known equations to needed ones is sometimes determined by physical arguments; sometimes it is purely a mathematical derivation.

3. The student should learn to approach any physical problem in an economical way, namely, from the general to the specific. It is not necessary to remember a vast amount of equations and relations applicable to special cases only. The clearer understanding of a physical theory would be negatively affected by too much unnecessary ballast.

[6]E. F. Taylor, J. A. Wheeler, *Space-Time Physics*, W. H. Freeman & Co., San Francisco, 1966.

4. All fundamental laws are simple because they express certain symmetries of nature. (The physical theory becomes more and more complicated when the laws are applied to larger and more complex systems, because the degree of symmetries decreases.) Usually, the fundamental laws are easily learned. The student needs to see however *how* these laws are applied to special cases. Problem solving, therefore, is a necessity. It helps the student master all steps from 1 to 4. The sentence "I really do understand the material—but I just cannot solve the problems" is self-contradictory.

QUESTIONS

1.1 The *modus tollens* (way of reasoning) in logic is an argument of the form:

> If *H* is true, then so is *I*.
> As evidence shows, *I* is not true.
> Therefore, *H* is not true.

The *fallacy of affirming the consequent* is an invalid argument of the form:

> If *H* is true, then so is *I*.
> As evidence shows, *I* is true.
> Therefore, *H* is true.

Discuss the importance of both arguments for the foundation of a scientific theory.

1.2 Comment on "Physics is too difficult for the physicist." (David Hilbert)

1.3 Ponder on "Science, like art, is not a copy of nature but a re-creation of her"[7] and on "A physical theory, being an amalgam of inventions, definitions, and laws, is regarded as a *model* for a certain part of nature, asserting not so much what nature *is*, but rather what it *is like*."[8]

For the following "common sense" questions state your answer within 5 seconds. Do not use force diagrams.

1.4 A rope is hanging over a pulley. A person attaches himself to one end of the rope and pulls on the other side to move himself up. Clearly, the force on the pulley is equal to the weight of the person (if the weights of the rope and of the pulley are neglected). The person now fastens the other end of the rope to the wall. Does the force on the pulley remain the same?

1.5 In 1654, Otto von Guericke, inventor of the air pump, demonstrated the existence of air pressure by evacuating two brass hemispheres and having two teams of eight horses pull on each side of the hemispheres. Assume that the air pressure inside the sphere is low enough that the hemispheres could just be pulled apart by the 16 horses. Assume now that the same air pressure exists as

---

[7]J. Bronowski, *Science and Human Values*, Harper & Row, New York, 1965.
[8]W. Rindler: *Essential Relativity*, Van Nostrand-Reinhold, New York, 1969.

before, but that now one of the hemispheres is fastened to a rigid wall. Then, eight horses cannot pull the hemispheres apart. True or false?

1.6 A wound up yo-yo is attached to a scale. The scale records the weight of the yo-yo. If this is released to unwind itself downward, the scale records a force (tension) that is smaller than the weight. Common sense would predict this result also. After passing through the lowest point the yo-yo is moving upward. Does the scale now indicate a force (tension) which is smaller than, equal to, or larger than the weight?

chapter

2 Mathematical Preliminaries

*Study without reflection is a waste of time,
reflection without study is dangerous.*

CONFUCIUS

2.1 VECTORS

It is presumed that the reader is familiar with scalars, vectors, and vector operations in three-dimensional space after completion of an introductory physics course. Vectors are denoted by boldface-type letters, **a**, **b**, **c**, etc. It is customary to use subscripts or superscripts such as $\bar{a}$, $\vec{a}$, $\tilde{a}$, or $\underset{\sim}{a}$ when vectors are written in longhand. The magnitudes of **a**, **b**, **c**, etc., are denoted by a, b, c, etc. A list of vector rules and operations is given in Appendix 1. The concept of vectors is not restricted to a three-dimensional space. Vectors may be defined in n-dimensional spaces also. Vectors in multidimensional spaces also appear in mechanics, for example, in problems dealing with oscillations of coupled systems.

The familiar definition of a vector as a quantity having magnitude and direction implies that a vector is independent of a coordinate system. (However, the coordinates of a vector are coordinate system dependent, see below.) Therefore, a vector remains unchanged under the operation of a parallel displacement. This means that if a certain observable quantity is a vector, a parallel displacement does not alter the meaning of the physical quantity. Velocity and acceleration may serve as examples.

There are observable quantities which, like vectors, have magnitude and direction but which cannot be moved parallel to themselves without changing their meaning. These quantities may be called vectors bound to a point, but they are not vectors in the strict sense. The radius " vector," defined as the vector from the origin O of a coordinate system to some point P in space, cannot be moved parallel to itself without simultaneously changing the points O and P. Another example is the torque about a given point of a rigid body.

It turns out that the relations and equations between vectors of each kind are coordinate system independent. This is a great advantage because all theoretical deductions from vector equations may be performed without reference to a specific coordinate system. Only when the vector equations are applied to a particular physical system in general must a coordinate system by introduced, in order to separate one vector equation into two or three coordinate equations that are then solved individually.

N vectors $\mathbf{e}_1, \mathbf{e}_2, \ldots \mathbf{e}_N$ are called linearly independent of each other if

$$a_1\mathbf{e}_1 + a_2\mathbf{e}_2 + \cdots + a_N\mathbf{e}_N = \sum_{i=1}^{N} a_i\mathbf{e}_i = 0$$

implies

$$a_1 = a_2 = \cdots = a_N = 0$$

Then the N vectors $\mathbf{e}_i$ span an N-dimensional vector space, and they can be chosen as the basis vectors in that space. The basis vectors, together with a fixed point (the origin) in space, define a coordinate system. In most cases the basis vectors are chosen at right angles to each other (orthogonal) and normalized to unit magnitude (unit vectors):

$$\mathbf{e}_i \cdot \mathbf{e}_j = \left\{ \begin{matrix} 0 & \text{if } i \neq j & \text{orthogonality} \\ 1 & \text{if } i = j & \text{unit magnitude} \end{matrix} \right\} \tag{2.1}$$

The Kronecker δ symbol is defined by

$$\delta_{ij} = \left\{ \begin{matrix} 0 & \text{if} & i \neq j \\ 1 & \text{if} & i = j \end{matrix} \right\} \tag{2.2}$$

Equation 2.1 then reads

$$\mathbf{e}_i \cdot \mathbf{e}_j = \delta_{ij} \tag{2.3}$$

for orthogonal and normalized basis vectors, or in short, orthonormal basis vectors. Sometimes (e.g., in crystallography), it is advantageous to use basis vectors that are not orthonormal.

If a coordinate system is introduced, then the (vector) components and the (scalar) coordinates of a vector $\mathbf{a}$ are uniquely determined. The following equivalent descriptions of a vector $\mathbf{a}$ are used when it is expressed in a coordinate system:

$$\mathbf{a} \equiv \mathbf{a}_1 + \mathbf{a}_2 + \cdots + \mathbf{a}_N \equiv a_1\mathbf{e}_1 + \cdots + a_N\mathbf{e}_N \tag{2.4}$$

If a certain set of basis vectors $\{\mathbf{e}_i\}$ is adopted, Eq. 2.4 may be abbreviated to

$$\mathbf{a} = (a_1, a_2, \ldots a_N)$$

It is convenient to introduce such a coordinate system relative to which certain coordinates of a given vector vanish and relative to which the equations of motion are in the simplest form.

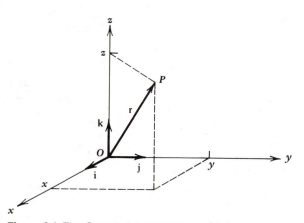

Figure 2.1 The Cartesian coordinate system.

2.2 SPECIAL COORDINATE SYSTEMS

2.2.1 The Cartesian Coordinate System

In a three-dimensional space the Cartesian, circular cylindrical, and spherical polar coordinate systems are of special importance. The Cartesian coordinate system is defined by the three orthonormal vectors

$$\mathbf{e}_1 = \mathbf{e}_x = \mathbf{i}, \quad \mathbf{e}_2 = \mathbf{e}_y = \mathbf{j} \quad \text{and} \quad \mathbf{e}_3 = \mathbf{e}_z = \mathbf{k} \tag{2.5}$$

fixed in space and forming a right-handed triad.[1] (See Figure 2.1.) A vector **a** may be expressed in one of the forms

$$\mathbf{a} = \mathbf{a}_x + \mathbf{a}_y + \mathbf{a}_z = a_x\mathbf{i} + a_y\mathbf{j} + a_z\mathbf{k} \tag{2.6}$$

or

$$a = (a_x, a_y, a_z)$$

The coordinates x, y, and z of a point P define the vector

$$\mathbf{r} = x\mathbf{i} + y\mathbf{j} + z\mathbf{k} \tag{2.7}$$

which is called the position or radius vector of the point P. The radius vector is a vector bound to a point, namely, to the origin O of the coordinate system.

2.2.2 The Circular Cylindrical Coordinate System

A point P in the three-dimensional space can be described by the coordinates ρ, φ, and z implicitly defined by

$$x = \rho \cos \varphi$$
$$y = \rho \sin \varphi \tag{2.8}$$
$$z = z,$$

---

[1]Right and left can be defined only in reference to a three-dimensional material object or process. Right and left cannot be defined in purely abstract terms.

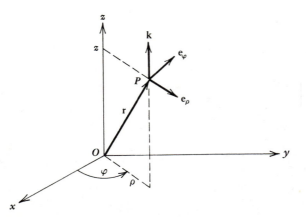

Figure 2.2 The circular cylindrical coordinate system.

or explicitly[2] by

$$\rho = \sqrt{x^2 + y^2}$$

$$\varphi = \frac{y}{|y|} \arctan \frac{y}{x} \tag{2.9}$$

$$z = z$$

ρ is the magnitude of the projection of $\mathbf{r}$ on the x-y plane, see Figure 2.2. The azimuthal angle φ is the angle between the vector $\mathbf{i}$ and the projection of $\mathbf{r}$ on the x-y plane. The vector $\mathbf{r}$ can then be written in the form

$$\mathbf{r} = x\mathbf{i} + y\mathbf{j} + z\mathbf{k} = \rho\cos\varphi\,\mathbf{i} + \rho\sin\varphi\,\mathbf{j} + z\mathbf{k}$$

$$= \rho(\cos\varphi\,\mathbf{i} + \sin\varphi\,\mathbf{j}) + z\mathbf{k}$$

$$= \rho\mathbf{e}_\rho + z\mathbf{k} \tag{2.10}$$

where

$$\mathbf{e}_\rho = \cos\varphi\,\mathbf{i} + \sin\varphi\,\mathbf{j} \tag{2.11}$$

is a unit vector in the x-y plane in the direction of the projection of $\mathbf{r}$. A third unit vector $\mathbf{e}_\varphi$ is introduced in such a way that $\mathbf{e}_1 = \mathbf{e}_\rho$, $\mathbf{e}_2 = \mathbf{e}_\varphi$, and $\mathbf{e}_3 = \mathbf{k}$ form a right-handed triad:

$$\mathbf{e}_\varphi = \mathbf{k} \times \mathbf{e}_\rho \tag{2.12}$$

In the Cartesian coordinate system $\mathbf{e}_\varphi$ is given by

$$\mathbf{e}_\varphi = -\sin\varphi\,\mathbf{i} + \cos\varphi\,\mathbf{j} \tag{2.13}$$

---

[2]We use the symbol "arc" for the inverse trigonometric functions to avoid possible confusion about the meaning of $\sin^{-1}x$, for example.

Observe that the unit vectors $\mathbf{e}_\rho$ and $\mathbf{e}_\varphi$ are no longer constant, but change their directions when the point P is moved perpendicular to $\mathbf{k}$. If $z = 0$, the circular cylindrical coordinate system reduces to the two-dimensional plane polar coordinate system.

2.2.3 The Spherical Polar Coordinate System

A point P in the three-dimensional space can be described by the coordinates r, θ, and φ implicitly defined by

$$x = r \sin \theta \cos \varphi$$

$$y = r \sin \theta \sin \varphi \qquad (2.14)$$

$$z = r \cos \theta$$

or explicitly by

$$r = \sqrt{x^2 + y^2 + z^2}$$

$$\theta = \arccos \frac{z}{\sqrt{x^2 + y^2 + z^2}} \qquad (2.15)$$

$$\varphi = \frac{y}{|y|} \arctan \frac{y}{x}$$

(See Figure 2.3.) Clearly, r is the magnitude of the vector $\mathbf{r}$. The polar angle θ is the angle between the vectors $\mathbf{k}$ and $\mathbf{r}$, whereas the azimuth angle φ is the angle between $\mathbf{i}$ and the projection of $\mathbf{r}$ on the x-y plane as in the case of circular cylindrical coordinates. In geometrical terms, the point P is defined by the intersection of a sphere of radius r, a cone of half angle θ about the z-axis, and a half plane bounded by the z-axis with azimuth angle φ.

Let $\mathbf{e}_1 = \mathbf{e}_r$ be the unit vector in the direction of $\mathbf{r}$. Then we have simply

$$\mathbf{r} = r\mathbf{e}_r \qquad (2.16)$$

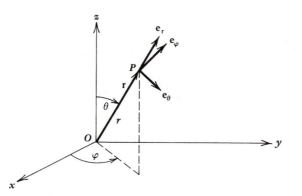

Figure 2.3 The spherical polar coordinate system.

Let $\mathbf{e}_2 = \mathbf{e}_\theta$ be a unit vector in the $\mathbf{k} - \mathbf{r}$ plane in the direction of increasing θ, and let $\mathbf{e}_3 = \mathbf{e}_\varphi$. Then the vectors $\mathbf{e}_r$, $\mathbf{e}_\theta$, and $\mathbf{e}_\varphi$ form a right-handed triad. In order to find the Cartesian coordinates of the unit vectors $\mathbf{e}_r$, $\mathbf{e}_\theta$, and $\mathbf{e}_\varphi$ it is convenient to first express these vectors in circular cylindrical coordinates. Inspecting Figure 2.3, and using Eqs. 2.11 and 2.12, we obtain

$$\mathbf{e}_r = \sin\theta\,\mathbf{e}_\rho + \cos\theta\,\mathbf{k}$$

$$= \sin\theta\cos\varphi\,\mathbf{i} + \sin\theta\sin\varphi\,\mathbf{j} + \cos\theta\,\mathbf{k}, \tag{2.17}$$

$$\mathbf{e}_\theta = \cos\theta\,\mathbf{e}_\rho - \sin\theta\,\mathbf{k}$$

$$= \cos\theta\cos\varphi\,\mathbf{i} + \cos\theta\sin\varphi\,\mathbf{j} - \sin\theta\,\mathbf{k}, \tag{2.18}$$

$$\mathbf{e}_\varphi = -\sin\varphi\,\mathbf{i} + \cos\varphi\,\mathbf{j} \tag{2.19}$$

All three basis vectors change when the point P is moved.

2.3 THE DERIVATIVE OF A VECTOR

2.3.1 Definition and Rules
A vector valued function $\mathbf{a}(x, y, z, t, \dots)$ is a vector that depends on the variables $x, y, z, t, \dots$.

$$\mathbf{a}(x, y, z, t, \dots) = \mathbf{a}_1(x, y, \dots) + \mathbf{a}_2(x, y, \dots) + \mathbf{a}_3(x, y, \dots)$$

$$= a_1(x, \dots)\mathbf{e}_1(x, \dots) + a_2(x, \dots)\mathbf{e}_2(x, \dots)$$

$$+ a_3(x, \dots)\mathbf{e}_3(x, \dots) \tag{2.20}$$

Let $\mathbf{a}$ be a vector valued function that depends on only one variable, say t. (Later, t will be identified with the time coordinate.) Then we have

$$\mathbf{a}(t) = \mathbf{a}_1(t) + \mathbf{a}_2(t) + \mathbf{a}_3(t)$$

$$= a_1(t)\mathbf{e}_1(t) + a_2(t)\mathbf{e}_2(t) + a_3(t)\mathbf{e}_3(t) \tag{2.21}$$

All of the rules of the calculus of differentiation also apply to vectors. The rate of change of $\mathbf{a}$ with respect to t is defined by

$$\frac{d\mathbf{a}}{dt} = \lim_{\Delta t \to 0} \frac{\mathbf{a}(t + \Delta t) - \mathbf{a}(t)}{\Delta t} \tag{2.22}$$

It is assumed that this limit exists. Because (2.22) is a vector equation, it must hold for all components $\mathbf{a}_i$ of $\mathbf{a}$:

$$\frac{d\mathbf{a}_i}{dt} = \lim_{\Delta t \to 0} \frac{\mathbf{a}_i(t + \Delta t) - \mathbf{a}_i(t)}{\Delta t} \tag{2.23}$$

Because of the product rule we have

$$\frac{d\mathbf{a}_i}{dt} = \frac{d}{dt}(a_i\mathbf{e}_i) = \frac{da_i}{dt}\mathbf{e}_i + a_i\frac{d\mathbf{e}_i}{dt}$$ (2.24)

da_i/dt is the rate of change of $\mathbf{a}_i$ in the direction of $\mathbf{e}_i$, and $d\mathbf{e}_i/dt$ is the rate of change of the vector $\mathbf{e}_i$. From $\mathbf{e}_i \cdot \mathbf{e}_i = 1$ follows that

$$\frac{d\mathbf{e}_i}{dt} \cdot \mathbf{e}_i + \mathbf{e}_i \cdot \frac{d\mathbf{e}_i}{dt} = 2\mathbf{e}_i \cdot \frac{d\mathbf{e}_i}{dt} = 0$$ (2.25)

Therefore, $d\mathbf{e}_i/dt$ is either equal to zero or it is orthogonal to $\mathbf{e}_i$.

2.3.2 Derivatives in Cartesian Coordinates
Because the basis vectors $\mathbf{i}$, $\mathbf{j}$, and $\mathbf{k}$ are constant in an inertial Cartesian coordinate system, we obtain

$$\frac{d\mathbf{a}}{dt} = \frac{da_x}{dt}\mathbf{i} + \frac{da_y}{dt}\mathbf{j} + \frac{da_z}{dt}\mathbf{k}$$ (2.26)

or shorter

$$\frac{d}{dt}(a_x, a_y, a_z) = \left(\frac{da_x}{dt}, \frac{da_y}{dt}, \frac{da_z}{dt}\right)$$ (2.27)

2.3.3 Derivatives in Circular Cylindrical Coordinates
The basis vectors $\mathbf{e}_i = \mathbf{e}_i(\mathbf{r}, t)$ in general vary in space and time, although in special cases they can depend on space or time alone. In the following chapters we need vectors that are attached to a particle (e.g., velocity, acceleration), and therefore we must construct a basis that is also attached to the particle. If the particle is followed, the independent variable $\mathbf{r}$ is then replaced by the time dependent location $\mathbf{r}(t)$, such that the basis vectors depend only on time.

In particular, we consider the circular cylindrical and the spherical polar coordinate systems with time dependent basis vectors. In order to find the derivatives of the coordinates of a vector in these coordinates we must make a detour via the Cartesian coordinate system. First we calculate the derivatives of $\mathbf{e}_\rho$ and $\mathbf{e}_\varphi$ using Eqs. 2.11 and 2.12:

$$\frac{d\mathbf{e}_\rho}{dt} = \frac{d}{dt}(\cos\varphi\,\mathbf{i} + \sin\varphi\,\mathbf{j}) = \frac{d\varphi}{dt}(-\sin\varphi\,\mathbf{i} + \cos\varphi\,\mathbf{j}) = \frac{d\varphi}{dt}\mathbf{e}_\varphi,$$ (2.28)

$$\frac{d\mathbf{e}_\varphi}{dt} = \frac{d}{dt}(-\sin\varphi\,\mathbf{i} + \cos\varphi\,\mathbf{j}) = \frac{d\varphi}{dt}(-\cos\varphi\,\mathbf{i} - \sin\varphi\,\mathbf{j}) = -\frac{d\varphi}{dt}\mathbf{e}_\rho$$ (2.29)

Then the derivative of an arbitrary vector $\mathbf{a}(t)$ expressed in circular cylindrical coordinates is found to be equal to

$$\frac{d\mathbf{a}}{dt} = \frac{d}{dt}(a_\rho\mathbf{e}_\rho) + \frac{d}{dt}(a_\varphi\mathbf{e}_\varphi) + \frac{d}{dt}(a_z\mathbf{k})$$

$$= \left(\frac{da_\rho}{dt} - a_\varphi\frac{d\varphi}{dt}\right)\mathbf{e}_\rho + \left(\frac{da_\varphi}{dt} + a_\rho\frac{d\varphi}{dt}\right)\mathbf{e}_\varphi + \frac{da_z}{dt}\mathbf{k}$$ (2.30)

or shorter

$$\frac{d}{dt}(a_\rho, a_\varphi, a_z) = \left(\frac{da_\rho}{dt} - a_\varphi\frac{d\varphi}{dt}, \frac{da_\varphi}{dt} + a_\rho\frac{d\varphi}{dt}, \frac{da_z}{dt}\right) \qquad (2.31)$$

2.3.4 Derivatives in Spherical Polar Coordinates
The derivatives of Eqs. 2.17, 2.18, and 2.19 are

$$\frac{d\mathbf{e}_r}{dt} = \left(\cos\theta\cos\varphi\frac{d\theta}{dt} - \sin\theta\sin\varphi\frac{d\varphi}{dt}\right)\mathbf{i}$$

$$+ \left(\cos\theta\sin\varphi\frac{d\theta}{dt} + \sin\theta\cos\varphi\frac{d\varphi}{dt}\right)\mathbf{j} - \sin\theta\frac{d\theta}{dt}\mathbf{k}$$

$$= \frac{d\theta}{dt}\mathbf{e}_\theta + \frac{d\varphi}{dt}\sin\theta\mathbf{e}_\varphi, \qquad (2.32)$$

$$\frac{d\mathbf{e}_\theta}{dt} = \left(-\sin\theta\cos\varphi\frac{d\theta}{dt} - \cos\theta\sin\varphi\frac{d\varphi}{dt}\right)\mathbf{i}$$

$$+ \left(-\sin\theta\sin\varphi\frac{d\theta}{dt} + \cos\theta\cos\varphi\frac{d\varphi}{dt}\right)\mathbf{j} - \cos\theta\frac{d\theta}{dt}\mathbf{k}$$

$$= -\frac{d\theta}{dt}\mathbf{e}_r + \frac{d\varphi}{dt}\cos\theta\mathbf{e}_\theta, \qquad (2.33)$$

$$\frac{d\mathbf{e}_\varphi}{dt} = \frac{d\varphi}{dt}(-\cos\varphi\mathbf{i} - \sin\varphi\mathbf{j})$$

$$= -\frac{d\varphi}{dt}\sin\theta\mathbf{e}_r - \frac{d\varphi}{dt}\cos\theta\mathbf{e}_\theta \qquad (2.34)$$

The last step in Eq. 2.34 is verified if one multiplies Eqs. 2.17 and 2.18 by $\sin\theta$ and $\cos\theta$, respectively, and then adds the two equations. The derivative of an arbitrary vector $\mathbf{a}$ in spherical polar coordinates is then given by

$$\frac{d\mathbf{a}}{dt} = \frac{d}{dt}(a_r\mathbf{e}_r) + \frac{d}{dt}(a_\theta\mathbf{e}_\theta) + \frac{d}{dt}(a_\varphi\mathbf{e}_\varphi) \qquad (2.35)$$

which yields because of Eqs. 2.32 through 2.34

$$\frac{d}{dt}(a_r, a_\theta, a_\varphi) = \left(\frac{da_r}{dt} - a_\theta\frac{d\theta}{dt} - a_\varphi\sin\theta\frac{d\varphi}{dt},\right.$$

$$\frac{da_\theta}{dt} + a_r\frac{d\theta}{dt} - a_\varphi\cos\theta\frac{d\varphi}{dt},$$

$$\left.\frac{da_\varphi}{dt} + a_r\sin\theta\frac{d\varphi}{dt} + a_\theta\cos\theta\frac{d\varphi}{dt}\right) \qquad (2.36)$$

The coordinates of higher order derivatives become increasingly more complex.

2.4 INTEGRATION OF A VECTOR WITH RESPECT TO ONE VARIABLE

If $\mathbf{a}(t)$ is a vector valued function of t, the indefinite integral of $\mathbf{a}$ over t is defined by

$$\int \mathbf{a}(t)\, dt = \mathbf{i}\int a_x(t)\, dt + \mathbf{j}\int a_y(t)\, dt + \mathbf{k}\int a_z(t)\, dt + \mathbf{C} \qquad (2.37)$$

where $\mathbf{C}$ is an arbitrary constant vector. The definite integral is similarly defined by

$$\int_{t_1}^{t_2} \mathbf{a}(t)\, dt = \mathbf{i}\int_{t_1}^{t_2} a_x(t)\, dt + \mathbf{j}\int_{t_1}^{t_2} a_y(t)\, dt + \mathbf{k}\int_{t_1}^{t_2} a_z(t)\, dt \qquad (2.38)$$

Integration in other coordinate systems cannot be performed so easily.

A word of caution is appropriate here. Although indefinite integrals are used in physics for the purpose of convenience, only definite integrals make sense in physics. Too often students regard the integral $\int_a^b f(x)\, dx$ as a function of x and try to differentiate the integral with respect to this dummy variable.

2.5 THE DEL OPERATOR

We consider a scalar function f, which depends on the three space coordinates x, y, z: $f = f(x, y, z) = f(\mathbf{r})$. The differential of f is equal to

$$df = \frac{\partial f}{\partial x}\, dx + \frac{\partial f}{\partial y}\, dy + \frac{\partial f}{\partial z}\, dz \qquad (2.39)$$

The right-hand side can be written as a scalar product of the infinitesimal displacement vector

$$d\mathbf{r} = (dx, dy, dz) \qquad (2.40)$$

and the vector

$$\text{grad } f = \left(\frac{\partial f}{\partial x}, \frac{\partial f}{\partial y}, \frac{\partial f}{\partial z} \right) \qquad (2.41)$$

which is called the gradient of f:

$$df = \text{grad } f \cdot d\mathbf{r} \qquad (2.42)$$

Because the form of Eq. 2.42 does not change if the scalar function $f(\mathbf{r})$ is replaced by another function $g(\mathbf{r})$, the vector property of the gradient must come from the differential operator

$$\nabla = \mathbf{i}\frac{\partial}{\partial x} + \mathbf{j}\frac{\partial}{\partial y} + \mathbf{k}\frac{\partial}{\partial z} = \left(\frac{\partial}{\partial x}, \frac{\partial}{\partial y}, \frac{\partial}{\partial z} \right) \qquad (2.43)$$

which is called the del or nabla operator. One treats ∇ formally like a vector, but one must remember that the differential operator has a meaning only when it is applied to a function of $\mathbf{r}$. Equation 2.42 becomes

$$df = \nabla f \cdot d\mathbf{r} \tag{2.44}$$

The interpretation of the vector grad $f \equiv \nabla f$ can easily be given. Let $ds = |d\mathbf{r}|$. Then from Eq. 2.44 follows

$$df = |\nabla f||d\mathbf{r}|\cos \sphericalangle(\nabla f, d\mathbf{r}) \tag{2.45}$$

or

$$\frac{df}{ds} = |\nabla f|\cos \sphericalangle(\nabla f, d\mathbf{r})$$

which is the projection of ∇f in the direction of $d\mathbf{r}$. The derivative df/ds has a maximum if ∇f is parallel to $d\mathbf{r}$. Therefore, ∇f indicates the direction and magnitude of the steepest slope of $f(\mathbf{r})$. If f is constant, that is, $df = 0$, along $d\mathbf{r}$, then either $\nabla f = 0$ or ∇f is perpendicular to $d\mathbf{r}$. The direction of the steepest slope is perpendicular to the surface $f(\mathbf{r}) = $ constant.

We briefly mention three other quantities related to the del operator. The curl of a vector $\mathbf{a}$ is the vector

$$\operatorname{curl} \mathbf{a} = \nabla \times \mathbf{a}$$

$$= \left(\frac{\partial a_z}{\partial y} - \frac{\partial a_y}{\partial z}\right)\mathbf{i} + \left(\frac{\partial a_x}{\partial z} - \frac{\partial a_z}{\partial x}\right)\mathbf{j} + \left(\frac{\partial a_y}{\partial x} - \frac{\partial a_x}{\partial y}\right)\mathbf{k} \tag{2.46}$$

The nonvanishing of curl a indicates that the vector $\mathbf{a}(r)$ has a rotational component. One also says that $\mathbf{a}$ possesses circulation. For example, the magnetic field $\mathbf{B}$ obeys $\nabla \times \mathbf{B} = \mu_0 \mathbf{J}$ where $\mathbf{J}$ is the electric current density vector. The divergence of a vector is the scalar

$$\operatorname{div} \mathbf{a} = \nabla \cdot \mathbf{a}$$

$$= \frac{\partial a_x}{\partial x} + \frac{\partial a_y}{\partial_y} + \frac{\partial a_z}{\partial z} \tag{2.47}$$

The divergence of a vector field $\mathbf{a}(\mathbf{r})$ may be the source of that field at the point $\mathbf{r}$. For example, the electric field $\mathbf{E}$ is related to the electric charge density ρ of all charges by $\varepsilon_0 \nabla \cdot \mathbf{E} = \rho$. The Laplacian ∇^2 is defined by

$$\nabla^2 = \nabla \cdot \nabla = \frac{\partial^2}{\partial x^2} + \frac{\partial^2}{\partial y^2} + \frac{\partial^2}{\partial z^2} \tag{2.48}$$

This operator plays a fundamental role in potential theory and wave theory.

2.6 LINE INTEGRALS

2.6.1 Line Integrals of Scalar Functions

We consider a function $f(\mathbf{r})$ defined on the three-dimensional space. Let C be a curve connecting two points P_1 and P_2 in the three-dimensional space, and let $ds = |d\mathbf{r}|$ denote the infinitesimal arc length of the curve C. The integral

$$\int_{P_1}^{P_2} f(\mathbf{r})\, ds \tag{2.49}$$
$$C$$

is called a line or path integral. It can be pictured in two dimensions as the area traced out by the curve C projected on the (surface) function $f(x, y)$ (see Figure 2.4). To evaluate the line integral, one introduces a parameter t that uniquely describes the curve C:

$$x = x(t), \quad y = y(t), \quad z = z(t)$$

The choice of a suitable parameter t usually depends on the shape of the curve. One may set t equal to the arc length s of the curve, $t = s$, or in two dimensions, one of the coordinates itself may be chosen as that parameter:

$$t = x: \quad y = y(x) \quad ds = \sqrt{1 + \left(\frac{dy}{dx}\right)^2}\, dx$$

The coordinates of the points P_1 and P_2 are determined by two values t_1 and t_2 of t:

$$P_1: \mathbf{r}_1 = \mathbf{r}(t_1), \quad P_2: \mathbf{r}_2 = \mathbf{r}(t_2)$$

The integral may then be written

$$\int_{P_1}^{P_2} f(\mathbf{r})\, ds = \int_{t_1}^{t_2} f(\mathbf{r}(t)) \frac{ds(t)}{dt}\, dt = \int_{t_1}^{t_2} g(t)\, dt$$
$$C$$

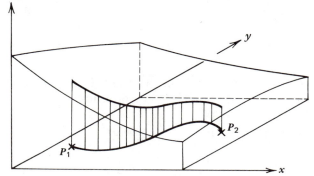

Figure 2.4 Line integral represented by an area.

where $g(t) = f\,ds/dt$. The value of the line integral is path dependent. This can be seen most easily if one sets $f(\mathbf{r}) = 1$. Then

$$\int_C f\,ds = \int_C ds = s_C$$

where s_C is the arc length of the curve C.

Example 2.1

Evaluate $\int_C xy\,ds$ where C is the section of a circle of radius R in the first quadrant.

We choose $t = \theta$. Then $x = R\cos\theta$, $y = R\sin\theta$, $ds = R\,d\theta$, $\theta_1 = 0$, and $\theta_2 = \pi/2$. The integral becomes

$$\int_C xy\,ds = \int_0^{\pi/2} R\cos\theta \cdot R\sin\theta \cdot R\,d\theta$$

$$= R^3 \int_0^{\pi/2} \cos\theta \sin\theta\,d\theta = \tfrac{1}{2}R^3 \qquad \blacksquare$$

2.6.2 Vector Line Integrals

A vector line integral is an integral of the form

$$\int_{P_1 \atop C}^{P_2} \mathbf{f}(\mathbf{r}) \cdot d\mathbf{r} \equiv \int_{P_1 \atop C}^{P_2} (f_x\,dx + f_y\,dy + f_z\,dz)$$

where $\mathbf{f}$ is a vector function and $d\mathbf{r}$ is the infinitesimal displacement vector along the curve C. If t is the path parameter, then

$$\int_{P_1 \atop C}^{P_2} \mathbf{f} \cdot d\mathbf{r} = \int_{P_1 \atop C}^{P_2} f\cos \sphericalangle(\mathbf{f}, d\mathbf{r})\,ds = \int_{t_1}^{t_2} f(t)\cos\theta(t)\frac{ds}{dt}\,dt = \int_{t_1}^{t_2} g(t)\,dt \quad (2.50)$$

and the integrand of the last integral is a function of t alone. It should be emphasized that the integral (2.50) is NOT equal to

$$\int_{x_1}^{x_2} f_x\,dx + \int_{y_1}^{y_2} f_y\,dy + \int_{z_1}^{z_2} f_z\,dz,$$

(see Problem 2.12).

Example 2.2

(a) Let $\mathbf{f} = a\mathbf{i} + bxy\mathbf{j}$. Let C_1 be the curve from the point $(0,0)$ to the point $(1,0)$ in x-direction and then to the point $(1,1)$ in y-direction.

$$\int_{C_1} \mathbf{f} \cdot d\mathbf{r} = \int_{x=0}^{x=1} \mathbf{f} \cdot d\mathbf{r}\Big|_{y=0} + \int_{y=0}^{y=1} \mathbf{f} \cdot d\mathbf{r}\Big|_{x=1}$$

$$= \int_0^1 a\,dx + \int_0^1 b \cdot 1 \cdot y\,dy = a + \tfrac{1}{2}b$$

(b) Let C_2 be the curve from the point $(0,0)$ to the point $(0,1)$ in y-direction and then to the point $(1,1)$ in x-direction

$$\int_{C_2} \mathbf{f} \cdot d\mathbf{r} = \int_{y=0}^{y=1} \mathbf{f} \cdot d\mathbf{r} \bigg|_{x=0} + \int_{x=0}^{x=1} \mathbf{f} \cdot d\mathbf{r} \bigg|_{y=1}$$

$$= \int_0^1 b \cdot 0 \cdot y \, dy + \int_0^1 a \, dx = a \qquad \blacksquare$$

2.6.3 Path Independent Vector Line Integrals

The last example has shown that vector line integrals are also in general path dependent. There exists, however, a very important class of vector functions for which the line integrals are path independent.

In order for the integral (2.50) to have the same value for all curves connecting the points P_1 and P_2, the integral can only depend on the coordinates $\mathbf{r}_1$ and $\mathbf{r}_2$ of the points P_1 and P_2, respectively:

$$\int_{P_1}^{P_2} \mathbf{f} \cdot d\mathbf{r} = \varphi(\mathbf{r}_2) - \varphi(\mathbf{r}_1)$$

This in turn is only possible if $\mathbf{f} \cdot d\mathbf{r}$ is a perfect differential of some function $\varphi(\mathbf{r})$:

$$d\varphi(\mathbf{r}) = \frac{\partial \varphi}{\partial x} dx + \frac{\partial \varphi}{\partial y} dy + \frac{\partial \varphi}{\partial z} dz = \nabla \varphi \cdot d\mathbf{r} = \mathbf{f} \cdot d\mathbf{r}$$

Because the last equation must be true for all displacement vectors $d\mathbf{r}$ at all points $\mathbf{r}$ in space, we find that

$$\mathbf{f} = \nabla \varphi \qquad (2.51)$$

or

$$f_x = \frac{\partial \varphi}{\partial x}, \quad f_y = \frac{\partial \varphi}{\partial y}, \quad f_z = \frac{\partial \varphi}{\partial z} \qquad (2.51\text{a})$$

from which follows

$$\frac{\partial f_x}{\partial y} = \frac{\partial f_y}{\partial x}, \quad \frac{\partial f_y}{\partial z} = \frac{\partial f_z}{\partial y}, \quad \frac{\partial f_z}{\partial x} = \frac{\partial f_x}{\partial z} \qquad (2.52)$$

or shorter

$$\nabla \times \mathbf{f} = 0 \qquad (2.52\text{a})$$

The statements

(a) $\int_{P_1}^{P_2} \mathbf{f} \cdot d\mathbf{r}$ is path independent
(b) $\oint \mathbf{f} \cdot d\mathbf{r} = 0$ for any closed path

(c) $\mathbf{f} \cdot d\mathbf{r}$ is a perfect differential, and

(d) $\nabla \times \mathbf{f} = 0$ everywhere

are equivalent.

Example 2.3

When is the line integral of $\mathbf{f} = ay\mathbf{i} + (x - z)\mathbf{j} + by\mathbf{k}$ path independent?
Applying condition (2.52) we obtain

$$\frac{\partial(ay)}{\partial y} = \frac{\partial(x-z)}{\partial x}, \quad \frac{\partial(x-z)}{\partial z} = \frac{\partial(by)}{\partial y}, \quad \frac{\partial(by)}{\partial x} = \frac{\partial(ay)}{\partial z}$$

or

$$a = 1, \quad -1 = b, \quad 0 = 0$$

Therefore, the line integral of the vector $y\mathbf{i} + (x - z)\mathbf{j} - y\mathbf{k}$ is path independent.

■

In order to find the scalar function φ from the vector $\mathbf{f}(\mathbf{r})$, one of the following methods may be applied:

1. The method of the indefinite line integral. One first evaluates the indefinite integrals

$$\varphi_1(x, y, z) = \int f_x(x, y, z)\, dx$$

$$\varphi_2(x, y, z) = \int f_y(x, y, z)\, dy$$

$$\varphi_3(x, y, z) = \int f_z(x, y, z)\, dz$$

Because

$$\frac{\partial \varphi_1}{\partial x} = f_x = \frac{\partial \varphi}{\partial x}, \quad \frac{\partial \varphi_2}{\partial y} = f_y = \frac{\partial \varphi}{\partial y}, \quad \frac{\partial \varphi_3}{\partial z} = f_z = \frac{\partial \varphi}{\partial z}$$

the integrals φ_i may differ from φ by integration "constants" C_i (or integration functions):

$$\varphi(x, y, z) = \varphi_1(x, y, z) + C_1(y, z) \tag{2.53a}$$

$$= \varphi_2(x, y, z) + C_2(x, z) \tag{2.53b}$$

$$= \varphi_3(x, y, z) + C_3(x, y) \tag{2.53c}$$

(Note that $\partial C_1/\partial x = \partial C_2/\partial y = \partial C_3/\partial z = 0$, which justifies the term *integration* constant although the C_i are functions of the coordinates.) If the integration "constants" are chosen in such a way that the right-hand sides of Eqs. 2.53a, b, and c are equal to each other, then φ is uniquely determined. Of course, a true constant C can still be added to the indefinite integral.

Example 2.4

What is the indefinite line integral φ of the vector given in Example 2.3?
The integrals φ_1, φ_2, and φ_3 are

$$\varphi_1 = \int f_x \, dx = \int y \, dx = xy$$

$$\varphi_2 = \int f_y \, dy = \int (x - z) \, dy = xy - yz$$

$$\varphi_3 = \int f_z \, dz = \int (-y) \, dy = -yz$$

The condition

$$\varphi = xy + C_1(y, z) + xy - yz + C_2(x, z) = -yz + C_3(x, y)$$

is satisfied if $C_1 = -yz$, $C_2 = 0$, and $C_3 = xy$. Therefore,

$$\varphi(x, y, z) = xy - yz + C \qquad \blacksquare$$

2. The method of the distinct path. Because it is known that the line integral is path independent, a distinct path may be chosen. If $\mathbf{f}(x_0, y_0, z_0) = 0$ for some point $\mathbf{r}_0$, one may select the path from (x_0, y_0, z_0) to (x, y_0, z_0) along the x-axis, then to (x, y, z_0) along the y-axis, and finally to the point (x, y, z) along the z-axis.

Example 2.5

Find φ for the vector $\mathbf{f} = y\mathbf{i} + (x - z)\mathbf{j} - y\mathbf{k}$ of Example 2.3. Because $f(0, 0, 0) = 0$, we have

$$\varphi(x, y, z) = \int_0^x f_x(x, 0, 0) \, dx + \int_0^y f_y(x, y, 0) \, dy + \int_0^z f_z(x, y, z) \, dz$$

$$= \int_0^x 0 \, dx + \int_0^y (x - 0) \, dy + \int_0^z (-y) \, dz$$

$$= xy - yz \qquad \blacksquare$$

2.7 COMPLEX NUMBERS

A complex number is a number of the form

$$z = a + ib$$

where a and b are real numbers and $i = \sqrt{-1}$ is the unit of imaginary numbers. $a = \text{Re}\{z\}$ is called the real part of z, $b = \text{Im}\{z\}$ is called the imaginary part of z. Any complex number can be represented by a point in the so-called complex plane (see Figure 2.5). The complex conjugate z^* of z is defined by

$$z^* = a - ib$$

(Most mathematicians and some physicists prefer the notation $\bar{z}$ to z^*.) Then

$$(z^*)^* = z,$$

$$a = \text{Re}\{z\} = \frac{1}{2}(z + z^*),$$

$$b = \text{Im}\{z\} = \frac{1}{2i}(z - z^*)$$

The magnitude of z, $|z| = r = \sqrt{a^2 + b^2} = \sqrt{zz^*}$, is called the modulus (or absolute value of z), the angle $\theta = (b/|b|) \arctan b/a$ is the argument of z. Then we can write z in the form

$$z = a + ib = r\cos\theta + ir\sin\theta$$

$$= r(\cos\theta + i\sin\theta) = re^{i\theta}$$

For the last step, we have used Euler's formula

$$e^{\pm i\theta} = \cos\theta \pm i\sin\theta \tag{2.54}$$

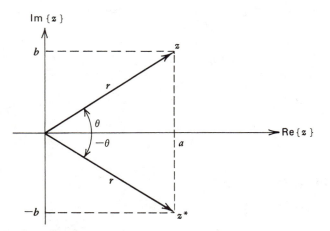

Figure 2.5 The complex plane.

which can be proved most easily with the help of the McLaurin series expansion

$$e^{i\theta} = 1 + \frac{1}{1!}(i\theta) + \frac{1}{2!}(i\theta)^2 + \frac{1}{3!}(i\theta)^3 + \cdots$$

$$= \left(1 - \frac{1}{2!}\theta^2 + \frac{1}{4!}\theta^4 - \cdots\right) + i\left(\frac{1}{1!}\theta - \frac{1}{3!}\theta^3 + \cdots\right)$$

$$= \cos\theta + i\sin\theta$$

For z^* we have the representation

$$z^* = re^{-i\theta}$$

If $r = 1$, z is equal to $e^{i\theta}$, and z is said to lie on the unit circle.

In classical physics, both the observable quantity (the quantity to be measured) and the measured numerical value of this quantity (the outcome of the measurement) are mathematically represented by the same function. And because the measured values are real numbers, this function must be real also. In quantum mechanics, the quantity to be measured and the outcome of the measurement are described by two different types of mathematical objects. The measured values are still real numbers, whereas the observable quantity is mapped on certain operators from which the allowed real observed values are calculated in a unique way.

Although the (classical) observable quantities are real functions, certain calculations may occasionally involve complex numbers. Students with little mathematical experience often consider the appearance of complex numbers as unreasonable or as a mysterious hocus pocus. They are reminded that complex numbers are convenient mathematical tools which can be used to solve equations in a quick and elegant way.

2.8 ELEMENTS OF MATRIX THEORY

An $m \times n$ matrix A is an array of numbers with m rows and n columns.

$$\mathbf{A} = (a_{ij}) = \begin{pmatrix} a_{11} & a_{12} & \cdots & a_{1n} \\ a_{21} & a_{22} & \cdots & a_{2n} \\ & & \vdots & \\ a_{m1} & a_{m2} & \cdots & a_{mn} \end{pmatrix}$$

i is called the row index, j is called the column index. Two matrices A and B are equal if $a_{ij} = b_{ij}$ for all i and j. Two $m \times n$ matrices A and B are added (subtracted) by adding (subtracting) the corresponding elements.

$$\mathbf{A} \pm \mathbf{B} = (a_{ij}) \pm (b_{ij}) = (a_{ij} \pm b_{ij}) \tag{2.55}$$

A scalar multiple of a matrix $\mathbf{A}$ is the matrix

$$c\mathbf{A} = (ca_{ij}),\tag{2.56}$$

that is, all elements of $\mathbf{A}$ are multiplied by the same scalar factor c. If $\mathbf{A}$ is an $m \times n$ matrix, and $\mathbf{B}$ is an $n \times p$ matrix, then an $m \times p$ matrix $\mathbf{C}$ can be constructed by the prescription

$$(c_{ik}) = \left(\sum_{j=1}^{n} a_{ij} b_{jk} \right)\tag{2.57}$$

$\mathbf{C}$ is called the matrix product of $\mathbf{A}$ and $\mathbf{B}$: $\mathbf{C} = \mathbf{AB}$. One says that the ith row of $\mathbf{A}$ is multiplied into the kth column of $\mathbf{B}$.

$$
\begin{pmatrix} c_{11} & c_{12} & \cdots & c_{1p} \\ c_{21} & c_{22} & \cdots & c_{2p} \\ \vdots & \vdots & & \vdots \\ c_{m1} & c_{m2} & \cdots & c_{mp} \end{pmatrix}
=
\begin{pmatrix} a_{11} & a_{12} & \cdots & a_{1n} \\ a_{21} & a_{22} & \cdots & a_{2n} \\ \vdots & \vdots & & \vdots \\ a_{m1} & a_{m2} & \cdots & a_{mn} \end{pmatrix}
\begin{pmatrix} b_{11} & b_{12} & \cdots & b_{1p} \\ b_{21} & b_{22} & \cdots & b_{2p} \\ \vdots & \vdots & & \vdots \\ b_{n1} & b_{n2} & \cdots & b_{np} \end{pmatrix}
$$

The element c_{12}, for example, is given by

$$c_{12} = a_{11}b_{12} + a_{12}b_{22} + a_{13}b_{32} + \cdots + a_{1n}b_{n2}.$$

Note that the numbers of columns of the matrix $\mathbf{A}$ must be equal to the numbers of rows of matrix $\mathbf{B}$.

Example 2.6

$$\begin{pmatrix} 1 & 0 & 2 \\ 0 & 1 & 1 \end{pmatrix} \begin{pmatrix} 1 & 1 & 2 \\ 1 & 1 & 0 \\ 0 & 1 & 0 \end{pmatrix} = \begin{pmatrix} 1 & 3 & 2 \\ 1 & 2 & 0 \end{pmatrix}\qquad ■$$

A matrix is called square if $m = n$, that is, the number of rows is equal to the numbers of columns. The elements a_{ii}, $i = 1 \ldots n$, of a square matrix form the main diagonal. For $n \times n$ matrices $\mathbf{A}$ and $\mathbf{B}$, both $\mathbf{AB}$ and $\mathbf{BA}$ are defined, but in general

$$\mathbf{AB} \neq \mathbf{BA}$$

Matrix multiplication is not commutative.

Example 2.7

$$\begin{pmatrix} 1 & 0 \\ 0 & 0 \end{pmatrix}\begin{pmatrix} 0 & 1 \\ 0 & 0 \end{pmatrix} = \begin{pmatrix} 0 & 1 \\ 0 & 0 \end{pmatrix} \quad \text{whereas} \quad \begin{pmatrix} 0 & 1 \\ 0 & 0 \end{pmatrix}\begin{pmatrix} 1 & 0 \\ 0 & 0 \end{pmatrix} = \begin{pmatrix} 0 & 0 \\ 0 & 0 \end{pmatrix}\qquad ■$$

The transpose A^T of a matrix A is the matrix

$$A^T = (a_{ij})^T = (a_{ji})$$

The rows and columns are interchanged. For square matrices, the elements of the main diagonal remain unchanged under the operation of transposition.

Example 2.8

$$\begin{pmatrix} 1 & 2 \\ 3 & 4 \end{pmatrix}^T = \begin{pmatrix} 1 & 3 \\ 2 & 4 \end{pmatrix}$$ ∎

Two subsequent transpositions are equivalent to no transposition.

$$(A^T)^T = A$$

The determinant of an $n \times n$ matrix is the number

$$\det A = |A| = \sum (\pm) a_{ij_1} a_{zj_2} \cdots a_{nj_n} \tag{2.58}$$

summed over all permutations $j_1 j_2 j_3 \cdots j_n$ of the set $\{1, 2, 3, \ldots n\}$. The sign is taken as $+$ or $-$ according to whether the permutation $j_1 j_2 j_3 \cdots j_n$ is even or odd. Thus, for $n = 2$ we have

$$\begin{vmatrix} a_{11} & a_{12} \\ a_{21} & a_{22} \end{vmatrix} = a_{11} a_{22} - a_{12} a_{21} \tag{2.58a}$$

whereas $n = 3$ leads to

$$\begin{vmatrix} a_{11} & a_{12} & a_{13} \\ a_{21} & a_{22} & a_{23} \\ a_{31} & a_{32} & a_{33} \end{vmatrix} = \begin{matrix} a_{11} a_{22} a_{33} + a_{12} a_{23} a_{31} + a_{13} a_{21} a_{32} \\ - a_{11} a_{23} a_{32} - a_{12} a_{21} a_{33} - a_{13} a_{22} a_{31} \end{matrix} \tag{2.58b}$$

We state without proof the following rules for $n \times n$ matrices:

$$(AB \cdots C)^T = C^T \cdots B^T A^T \tag{2.59}$$

$$|AB \cdots C| = |A||B| \cdots |C| \tag{2.60}$$

$$|A^T| = |A| \tag{2.61}$$

$$|-A| = (-1)^n |A| \tag{2.62}$$

We are now prepared to define several important types of matrices. A diagonal matrix is a matrix for which all off-diagonal elements vanish,

$$a_{ij} = 0, \quad i \neq j$$

The determinant of a diagonal matrix is equal to $a_{11} a_{22} \cdots a_{nn}$.

Example 2.9

The matrix $\begin{pmatrix} 3 & 0 & 0 \\ 0 & -1 & 0 \\ 0 & 0 & 0 \end{pmatrix}$ is diagonal. ■

The unit matrix **1** (or identity matrix) is the matrix $\mathbf{1} = (\delta_{ij})$. The determinant of **1** is equal to 1.

Example 2.10

The unit matrix in a three-dimensional space is $\begin{pmatrix} 1 & 0 & 0 \\ 0 & 1 & 0 \\ 0 & 0 & 1 \end{pmatrix}$ ■

A symmetric matrix is defined by the condition

$$\mathbf{A} = \mathbf{A}^T \tag{2.63}$$

or equivalently

$$(a_{ij}) = (a_{ji})$$

Example 2.11

The matrix $\begin{pmatrix} 2 & -3 & 1 \\ -3 & 4 & 0 \\ 1 & 0 & 3 \end{pmatrix}$ is symmetric. ■

An antisymmetric (or skew symmetric) matrix is defined by

$$\mathbf{A} = -\mathbf{A}^T \tag{2.64}$$

or equivalently

$$(a_{ij}) = -(a_{ji})$$

The diagonal elements of an antisymmetric matrix are zero.

Example 2.12

The matrix $\begin{pmatrix} 0 & -2 & 3 \\ 2 & 0 & 1 \\ -3 & -1 & 0 \end{pmatrix}$ is antisymmetric. ■

A matrix is called singular if its determinant is equal to zero:

$$|\mathbf{A}| = 0 \tag{2.65}$$

If A is nonsingular, that is, if $|\mathbf{A}| \neq 0$, a matrix $\mathbf{A}^{-1}$ can be found such that

$$\mathbf{A}^{-1}\mathbf{A} = \mathbf{A}\mathbf{A}^{-1} = \mathbf{1} \tag{2.66}$$

$\mathbf{A}^{-1}$ is called the inverse of A. We need not discuss the lengthy procedure for calculating the elements of the matrix A^{-1} from A. Fortunately, we need the inverse of orthogonal matrices only, for which the elements of A can easily be found by transposition.

A matrix is called orthogonal if

$$A^{-1} = A^T \tag{2.67}$$

or equivalently,

$$A^T A = AA^T = 1 \tag{2.68}$$

Applying Eqs. 2.60 and 2.61 to 2.68, we obtain

$$|A^T A| = |A^T||A| = |A|^2 = 1$$

from which follows

$$|A| = \pm 1 \tag{2.69}$$

for an orthogonal matrix.

Example 2.13

The matrix $A = \begin{pmatrix} \cos\varphi & \sin\varphi & 0 \\ -\sin\varphi & \cos\varphi & 0 \\ 0 & 0 & 1 \end{pmatrix}$ is orthogonal because

$$AA^T = \begin{pmatrix} \cos\varphi & \sin\varphi & 0 \\ -\sin\varphi & \cos\varphi & 0 \\ 0 & 0 & 1 \end{pmatrix} \begin{pmatrix} \cos\varphi & -\sin\varphi & 0 \\ \sin\varphi & \cos\varphi & 0 \\ 0 & 0 & 1 \end{pmatrix} = 1 \qquad \blacksquare$$

The coordinates of a vector relative to an orthonormal basis can be written in the form of an $n \times 1$ matrix,

$$\mathbf{x} \rightarrow X = \begin{pmatrix} x_1 \\ x_2 \\ \vdots \\ x_n \end{pmatrix} \tag{2.70}$$

Its transpose is the $1 \times n$ matrix

$$X^T = \begin{pmatrix} x_1 \\ x_2 \\ \vdots \\ x_n \end{pmatrix}^T = (x_1, x_2, \ldots, x_n) \tag{2.71}$$

The inner product of two vectors $\mathbf{x}$ and $\mathbf{y}$ becomes the matrix product

$$X^T Y = \mathbf{x} \cdot \mathbf{y} = \mathbf{y} \cdot \mathbf{x} = Y^T X \tag{2.72}$$

because

$$(x_1, x_2, \ldots x_n) \begin{pmatrix} y_1 \\ y_2 \\ \vdots \\ y_n \end{pmatrix} = x_1 y_1 + x_2 y_2 + \cdots x_n y_n$$

An $n \times n$ matrix A transforms a column matrix (vector) into another column matrix (vector):

$$AX = Y \tag{2.73}$$

which is explicitly

$$\begin{pmatrix} a_{11} & a_{12} & \cdots & a_{1n} \\ a_{21} & a_{22} & \cdots & a_{2n} \\ \vdots & \vdots & & \vdots \\ a_{n1} & a_{n2} & \cdots & a_{nn} \end{pmatrix} \begin{pmatrix} x_1 \\ x_2 \\ \vdots \\ x_n \end{pmatrix} = \begin{pmatrix} a_{11}x_1 + a_{12}x_2 + \cdots a_{1n}x_n \\ a_{21}x_1 + a_{22}x_2 + \cdots a_{2n}x_n \\ \vdots \\ a_{n1}x_1 + a_{n2}x_2 + \cdots a_{nn}x_n \end{pmatrix} = \begin{pmatrix} y_1 \\ y_2 \\ \vdots \\ y_n \end{pmatrix}$$

In particular, for the unit matrix 1 we obtain

$$1X = X \tag{2.74}$$

2.9 COORDINATE TRANSFORMATIONS

We restrict ourselves to transformations in a three-dimensional space. Most of the equations we shall derive apply to n-dimensional spaces also. Let S and S' be two coordinate systems defined by the orthonormal bases $\{e_1, e_2, e_3\}$ and $\{e_1', e_2', e_3'\}$ respectively (see Figure 2.6).

Any vector x can be decomposed in the following two ways:

$$x = x_1 e_1 + x_2 e_2 + x_3 e_3 = \sum_{j=1}^{3} x_j e_j$$

$$= x_1' e_1' + x_2' e_2' + x_3' e_3' = \sum_{i=1}^{3} x_i' e_i' \tag{2.75}$$

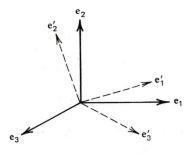

Figure 2.6 Rotated basis vectors.

We wish to find a relation between the coordinates x_1', x_2', x_3' of $\mathbf{x}$ in S' and the coordinates x_1, x_2, x_3 of $\mathbf{x}$ in S. For this purpose, let us multiply Eq. 2.75 by $\mathbf{e}_i'$.

$$\mathbf{e}_i' \cdot \mathbf{x} = x_i' = \sum_{j=1}^{3} \mathbf{e}_i' \cdot \mathbf{e}_j x_j \tag{2.76}$$

The inner products $\mathbf{e}_i' \cdot \mathbf{e}_j = \cos \sphericalangle(\mathbf{e}_i', \mathbf{e}_j)$, which are the direction cosines of the angles between $\mathbf{e}_i'$ and $\mathbf{e}_j$, define a 3×3 matrix R:

$$\mathsf{R} = (r_{ij}) = (\mathbf{e}_i' \cdot \mathbf{e}_j) \tag{2.77}$$

Equation 2.76 then goes over into

$$x_i' = \sum_{j=1}^{3} r_{ij} x_j \tag{2.78}$$

which is in matrix form

$$\mathsf{X}' = \mathsf{R}\mathsf{X} \tag{2.79}$$

The matrix R transforms the coordinates x_1, x_2, x_3 into the coordinates x_1', x_2', x_3'.

In order to find the inverse transformation from the primed to the unprimed coordinates, we multiply Eq. 2.75 by $\mathbf{e}_j$. Then

$$\mathbf{x} \cdot \mathbf{e}_j = x_j = \sum_{i=1}^{3} x_i' \mathbf{e}_i' \cdot \mathbf{e}_j$$

$$= \sum_{i=1}^{3} x_i' r_{ij}$$

$$= \sum_{i=1}^{3} r_{ji}^T x_i' \tag{2.80}$$

or in matrix form

$$\mathsf{X} = \mathsf{R}^T \mathsf{X}' \tag{2.81}$$

Inserting X' from Eq. 2.79 into the last equation we obtain

$$\mathsf{X} = \mathsf{R}^T \mathsf{R}\mathsf{X} \tag{2.82}$$

Because this equation is valid for all vectors X, we conclude that

$$\mathsf{R}^T \mathsf{R} = 1 \quad \text{or} \quad \mathsf{R}^T = \mathsf{R}^{-1}$$

Any matrix R which transforms the coordinates of a vector relative to an orthonormal basis into the coordinates relative to another orthonormal basis is orthogonal.

Example 2.14

What is the matrix R for the transformation from the Cartesian basis $\{e_1, e_2, e_3\} = \{i, j, k\}$ to the circular cylindrical basis $\{e'_1, e'_2, e'_3\} = \{e_\rho, e_\varphi, k\}$? Using (2.11), (2.12), and (2.13), we obtain

$$e'_1 = \cos\varphi e_1 + \sin\varphi e_2, \quad e'_2 = -\sin\varphi e_1 + \cos\varphi e_2, \quad e'_3 = e_3$$

Therefore,

$$r_{11} = e'_1 \cdot e_1 = \cos\varphi \qquad r_{12} = e'_1 \cdot e_2 = \sin\varphi \qquad r_{13} = e'_1 \cdot e_3 = 0$$

$$r_{21} = e'_2 \cdot e_1 = -\sin\varphi \qquad r_{22} = e'_2 \cdot e_2 = \cos\varphi \qquad r_{23} = e'_2 \cdot e_3 = 0$$

$$r_{31} = e'_3 \cdot e_1 = 0 \qquad r_{32} = e'_3 \cdot e_2 = 0 \qquad r_{33} = e'_3 \cdot e_3 = 1$$

and

$$R = \begin{pmatrix} \cos\varphi & \sin\varphi & 0 \\ -\sin\varphi & \cos\varphi & 0 \\ 0 & 0 & 1 \end{pmatrix} \tag{2.83}$$

The coordinates of the vector x in circular cylindrical coordinates are, therefore,

$$\begin{pmatrix} x'_1 \\ x'_2 \\ x'_3 \end{pmatrix} = RX = \begin{pmatrix} (\cos\varphi x_1 + \sin\varphi x_2) \\ (-\sin\varphi x_1 + \cos\varphi x_2) \\ x_3 \end{pmatrix} \tag{2.84}$$

∎

An orthogonal coordinate transformation preserves the magnitude of a vector X, because if

$$Y = AX,$$

then

$$Y^T = (AX)^T = X^T A^T$$

and

$$Y^T Y = (X^T A^T)(AX) = X^T A^T A X = X^T 1 X = X^T X$$

There are two possibilities. If $\det A = +1$, the transformation is a rotation. If $\det A = -1$, the transformation is a reflection of *one* or *three* coordinate axes or a rotation and a reflection combined (sometimes called reflexotation or rotoreflection). A reflection of *two* coordinate axes is equivalent to a rotation of π about the third axis.

Example 2.15

The matrix R, Eq. 2.83, describes a rotation in the x-y plane.

$$\det R = \cos^2 \varphi - (-\sin^2 \varphi) = 1$$

If $\varphi = \pi$, then

$$R = \begin{pmatrix} -1 & 0 & 0 \\ 0 & -1 & 0 \\ 0 & 0 & 1 \end{pmatrix}$$

which represents a reflection of the x- and y-axes. ■

Any transformation matrix can be used to describe two different but closely related transformations.

1. In the case of a *coordinate transformation* the basis vectors are transformed, which causes the coordinates of a point P fixed in space to change.
2. In the case of a *point transformation* a point P in space is transformed into another point P', whereas the basis vectors remain unchanged. We illustrate these two transformations in the following example.

Example 2.16

Let $\mathbf{r} = (x, y, 0)$ be the radius vector of a point P in the x-y plane, and let

$$R(\theta) = R = \begin{pmatrix} \cos\theta & \sin\theta & 0 \\ -\sin\theta & \cos\theta & 0 \\ 0 & 0 & 1 \end{pmatrix}$$

be the rotation matrix for a rotation by θ about the z-axis.

1. Applying (2.78) to the coordinates of the radius vector $\mathbf{r}$ of the point P, we obtain the new coordinates

$$x' = \cos\theta x + \sin\theta y$$

$$y' = -\sin\theta x + \cos\theta y$$

as shown in Figure 2.7. The angle between the vectors $\mathbf{r}$ and $\mathbf{i}'$ decreases.

2. We wish to rotate the point P (vector $\mathbf{r}$) by the same angle θ. The angle between $\mathbf{r}$ and $\mathbf{i}$ increases in this case, and we must apply to $\mathbf{r}$ the matrix $R(-\theta) = R^{-1}(\theta) = R^T(\theta)$. The coordinates of the new point are

$$x' = \cos(-\theta)x + \sin(-\theta)y = \cos\theta x - \sin\theta y$$

$$y' = -\sin(-\theta)x + \cos(-\theta)y = \sin\theta x + \cos\theta y$$

as shown in Figure 2.8. ■

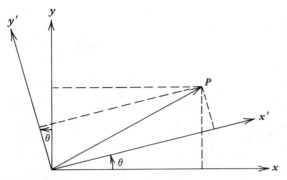

Figure 2.7 Coordinate transformation.

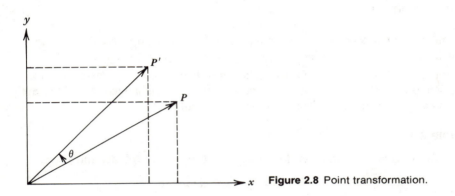

Figure 2.8 Point transformation.

2.10 TENSORS

Let dx_j be the coordinates of an infinitesimal displacement relative to the coordinate system S, and let dx_i' be the coordinates of the same infinitesimal displacement relative to the coordinate system S'. If the x_i' are considered as functions of the x_j, then we have

$$dx_i' = \sum_{j=1}^{N} \frac{\partial x_i'}{\partial x_j} dx_j, \quad i = 1, 2, \ldots N \tag{2.85}$$

or

$$dx_i' = \sum_{j=1}^{N} A_{ij} dx_j$$

where

$$A_{ij} = \frac{\partial x_i'}{\partial x_j} \tag{2.86}$$

is the matrix that transforms the dx_j into the dx_i'. Because

$$\delta_{ik} = \frac{\partial x_i'}{\partial x_k'} = \sum_{j=1}^{N} \frac{\partial x_i'}{\partial x_j} \frac{\partial x_j}{\partial x_k'} = \sum_{j=1}^{N} A_{ij} \frac{\partial x_j}{\partial x_k'}$$

it follows that

$$\frac{\partial x_j}{\partial x_k'} = \left(A_{jk} \right)^{-1}$$

Any set of quantities $M_{j_1 j_2 \cdots j_n}$ in S, which transforms according to the rule

$$M_{i_1 \ldots i_n}'(x_i') = \sum_{j_1} \cdots \sum_{j_n} A_{i_1 j_1} \cdots A_{i_n j_n} M_{j_1 \ldots j_n}(x_j) \qquad (2.87)$$

into another set in S', where $A_{i_k j_k}$ is given by (2.86), is said to be a contravariant tensor of rank n under the given transformation. A contravariant tensor of rank $n = 1$ is called a contravariant vector. It transforms like the coordinates of the infinitesimal displacement (2.85):

$$V_i' = \sum_j A_{ij} V_j$$

Of special importance are also tensors of rank two. The transformation rule

$$M_{ij}' = \sum_k \sum_l A_{ik} A_{jl} M_{kl}$$

$$= \sum_k \sum_l A_{ik} M_{kl} A_{lj}^T \qquad (2.88)$$

can conveniently be written as a product of matrices,

$$\mathbf{M}' = \mathbf{A} \mathbf{M} \mathbf{A}^T \qquad (2.89)$$

Depending on whether the matrix (M_{kl}) is symmetric or antisymmetric the tensor is called symmetric or antisymmetric. Although tensors of rank two can be written in the form of square matrices, square matrices are not necessarily tensors. In order for a mathematical object $\mathbf{M}$ to be a tensor, it is necessary for it to have the transformation property (2.87). It is possible that $\mathbf{M}$ is a tensor under one transformation T_1 (e.g., a rotation) whereas it is not simultaneously a tensor under another transformation T_2 (e.g., a reflection).

A tensor of rank zero is called a scalar. A scalar carries no indices. A scalar is invariant (i.e., it remains unchanged) under the transformation, and we have

$$M' = M'(x_i') = M(x_j) = M$$

The coordinates $\partial/\partial x_j$ of the del operator do not form a contravariant vector because

$$\frac{\partial M'}{\partial x_i'} = \frac{\partial M}{\partial x_i'} = \sum_j \frac{\partial M}{\partial x_j}\frac{\partial x_j}{\partial x_i'} = \sum_j \frac{\partial M}{\partial x_j}\left(A_{ji}\right)^{-1}$$

and hence,

$$\frac{\partial}{\partial x_i'} = \sum_j \left(A_{ij}\right)^{-1T}\frac{\partial}{\partial x_j}$$

Quantities that transform like the $\partial/\partial x_i$ are called covariant vectors. Covariant tensors of rank n can also be defined by analogy to (2.87). The difference between co- and contravariant tensors disappears if the coefficients of A form an orthogonal matrix, because then $(A^{-1})^T = A$. This is the case, for example, if the transformation leads from one Cartesian coordinate system into another Cartesian coordinate system. For further information about tensors consult the special literature.[3]

2.11 POLAR AND AXIAL VECTORS

We distinguish two types of vectors in the three-dimensional space.

A polar vector is a quantity with three components which, relative to a given coordinate system, change the sign when all basis vectors are reflected. If a_1, a_2, a_3 are the coordinates of a polar vector $\mathbf{a}$ in the coordinate system S, then

$$(a_1', a_2', a_3') = (-a_1, -a_2, -a_3)$$

are the coordinates of $\mathbf{a}$ in the reflected system S'. This means in tensor language that a polar vector is a tensor of rank one under reflection. Because the matrix describing the reflection of all coordinate axes is given by

$$A = -1 = -\left(\delta_{ij}\right)$$

we have

$$x_i' = \sum_{j=1}^{3} A_{ij}x_j = -\sum_{j=1}^{3} \delta_{ij}x_j = -x_i$$

A polar vector is graphically represented by an arrow. The direction of the arrow indicates the direction of the vector, and the length of the arrow corresponds to the magnitude of the vector (see Figures 2.9a and 2.9b). Examples of polar vectors are displacement, velocity, and force.

---

[3]For example, J. Mathews, R. L. Walker: *Mathematical Methods of Physics*, W. A. Benjamin Inc., New York, 1965; D. Lovelock, H. Rund: *Tensors, Differential Forms and Variational Principles*, Wiley-Interscience, New York, 1975; J. A. Schouten, *Ricci Calculus*, Springer, Berlin, 1954.

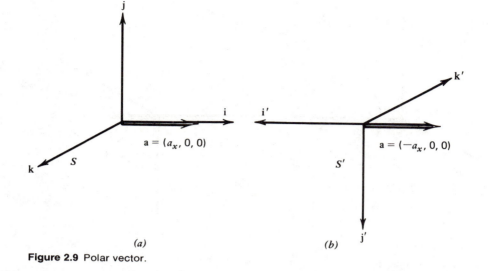

$a = (a_x, 0, 0)$

$a = (-a_x, 0, 0)$

(a)

(b)

Figure 2.9 Polar vector.

An axial vector is a quantity with three components which relative to a given coordinate system do not change the sign when all basis vectors are reflected. An axial vector cannot be represented by a tensor of rank one; however, it possesses all the properties of an antisymmetric tensor of rank two which has three independent components in the three-dimensional space. Thus, if the axial vector is represented by

$$\mathbf{M} = \begin{pmatrix} 0 & a_z & -a_y \\ -a_z & 0 & a_x \\ a_y & -a_x & 0 \end{pmatrix},$$

then the transformation rule (2.89) implies that

$$\mathbf{M}' = \mathbf{AMA}^T = (-1)\mathbf{M}(-1) = \mathbf{M},$$

and the components of $\mathbf{M}$ do not change sign under the reflection operation. Why, then, is this mathematical object called an axial vector? Because there is a one-to-one correspondence between the elements of any 3×3 antisymmetric tensor $\mathbf{M}$ and the coordinates of a vector $\mathbf{m}$ given by

$$m_1 = M_{23}, \quad m_2 = M_{31}, \quad \text{and} \quad m_3 = M_{12}$$

if we disregard the transformation properties under the reflection. This is why the axial vector can be thought of as an arrow provided with a screw sense (sense of rotation), whereby the direction of the arrow represents both the axis of rotation and the sense of rotation defined by the right-hand rule, and the length of the

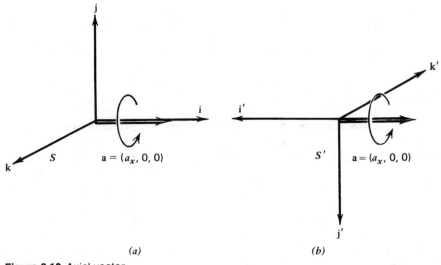

(a)   *(b)*

Figure 2.10 Axial vector.

arrow corresponds to the magnitude of the axial vector. The picture of an arrow, provided with a screw sense, together with the fact that vectors and matrices obey the same algebraic rules for addition and multiplication by scalars allow us to treat axial vectors in the same way as polar vectors. We therefore use the same symbol to denote both types of vectors.

In the example shown in Figures 2.10a and 2.10b, the rotation axis is the x-axis. The sense of rotation from the positive y-axis to the positive z-axis remains unchanged when the coordinate system S is transformed into the reflected system S'. Examples of axial vectors are angular velocity, angular momentum, and torque.

2.12 THE EIGENVALUE PROBLEM

Let A be any given $n \times n$ matrix. We ask the following question: Do vectors X exist such that

$$AX = \lambda X \qquad (2.90)$$

where λ is some scalar, that is, is it possible to find a vector X which is transformed under A into another vector parallel or antiparallel to X? Equation 2.90, which can be written in the form

$$(A - \lambda 1)X = 0 \qquad (2.91)$$

represents a system of n homogeneous linear equations in the n unknowns x_k. It is shown in linear algebra that such a system has nontrivial solutions $X \neq 0$ if and only if

$$
\det(A - \lambda \mathbb{1}) = \begin{vmatrix} a_{11} - \lambda & a_{12} & \cdots & a_{1n} \\ a_{21} & a_{22} - \lambda & \cdots & a_{2n} \\ \vdots & \vdots & & \vdots \\ a_{n1} & a_{n2} & \cdots & a_{nn} - \lambda \end{vmatrix} = 0 \qquad (2.92)
$$

The determinant yields a polynomial of degree n for λ:

$$
(-\lambda)^n + a_{n-1}(-\lambda)^{n-1} + \cdots - a_1\lambda + a_0 = 0 \qquad (2.93)
$$

which has n solutions $\lambda_1, \lambda_2, \ldots \lambda_n$ by the fundamental theorem of algebra. The λ_i may be real or complex numbers, and they may or may not be different from each other. Because there are only n values λ_i, we may find n vectors $X^{(i)}$ satisfying the condition (2.90). λ_i is called an eigenvalue[4] of A. Likewise, $X^{(i)}$ is called an eigenvector of A belonging to λ_i. Because Eq. 2.91 is homogeneous, any multiple of $X^{(i)}$ is also an eigenvector. One often requires $X^{(i)}$ to be a unit vector,

$$
X^{(i)T}X^{(i)} = 1
$$

and in this case one says that the eigenvector has been normalized to one.

Example 2.17

Find the eigenvalues of $A = \begin{pmatrix} 2 & 1 \\ 1 & 2 \end{pmatrix}$.

Evaluating the determinant (2.92) we obtain

$$
\det(A - \lambda \mathbb{1}) = \begin{vmatrix} 2 - \lambda & 1 \\ 1 & 2 - \lambda \end{vmatrix} = (2 - \lambda)^2 - 1 = 0
$$

from which follows that

$$
\lambda_1 = 1, \quad \lambda_2 = 3 \qquad \blacksquare
$$

To find the eigenvectors associated with the λ_i, one must solve the n systems of equations

$$
(A - \lambda_i \mathbb{1})X^{(i)} = 0, \quad i = 1, 2, \ldots n \qquad (2.94)
$$

for the coordinates $x_l^{(i)}, \quad l = 1, 2, \ldots n$.

Example 2.18

Find the eigenvectors $X^{(1)}$ and $X^{(2)}$ for the matrix given in Example 2.17. For $\lambda_1 = 1$ we obtain

$$
(A - \lambda_1 \mathbb{1})X^{(1)} = 0
$$

[4]This combination of a German and an English word has now become generally accepted (eigen = proper, belonging to, characteristic).

or

$$\begin{pmatrix} 2-1 & 1 \\ 1 & 2-1 \end{pmatrix} \begin{pmatrix} x_1^{(1)} \\ x_2^{(1)} \end{pmatrix} = \begin{pmatrix} x_1^{(1)} + x_2^{(1)} \\ x_1^{(1)} + x_2^{(1)} \end{pmatrix} = 0$$

Setting $c = x_1^{(1)} = -x_2^{(1)}$ we have

$$\mathbf{X}^{(1)} = \begin{pmatrix} c \\ -c \end{pmatrix} = c \begin{pmatrix} 1 \\ -1 \end{pmatrix}$$

The normalization condition $\mathbf{X}^{(1)T}\mathbf{X}^{(1)} = x_1^{(1)}x_1^{(1)} + x_2^{(1)}x_2^{(1)} = 1$ implies that $c = 1/\sqrt{2}$. Then the normalized eigenvector $X^{(1)}$ belonging to $\lambda_1 = 1$ is

$$\mathbf{X}^{(1)} = \frac{1}{\sqrt{2}} \begin{pmatrix} 1 \\ -1 \end{pmatrix}$$

Similarly, one obtains

$$\mathbf{X}^{(2)} = \frac{1}{\sqrt{2}} \begin{pmatrix} 1 \\ 1 \end{pmatrix}$$

for the eigenvector belonging to $\lambda_2 = 3$. ∎

The following two theorems are of special importance.

1   The eigenvalues of a real symmetric matrix are real.
2   If two eigenvalues of a symmetric matrix are different from each other, the corresponding eigenvectors are orthogonal.

Proof of 1   We assume that both λ_i and $X_k^{(i)}$ are complex numbers. The eigenvalue equation

$$\mathbf{AX}^{(i)} = \lambda_i \mathbf{X}^{(i)} \tag{2.95}$$

implies

$$\mathbf{AX}^{(i)*} = \lambda_i^* \mathbf{X}^{(i)*} \tag{2.96}$$

We multiply Eq. 2.95 from the left by $\mathbf{X}^{(i)*T}$. Then

$$\mathbf{X}^{(i)*T}\mathbf{AX}^{(i)} = \lambda_i \mathbf{X}^{(i)*T}\mathbf{X}^{(i)} \tag{2.97}$$

The left-hand side is, because of Eqs. 2.59 and 2.96, equal to

$$(\mathbf{A}^T\mathbf{X}^{(i)*})^T\mathbf{X}^{(i)} = (\mathbf{AX}^{(i)*})^T\mathbf{X}^{(i)} = \lambda_i^*\mathbf{X}^{(i)*T}\mathbf{X}^{(i)}$$

Therefore,

$$(\lambda_i - \lambda_i^*)\mathbf{X}^{(i)*T}\mathbf{X}^{(i)} = 0$$

Because $X^{(i)*T}X^{(i)}$ is real and > 0, it follows that $\lambda_i = \lambda_i^*$. Therefore, λ_i is a real number.

Proof of 2 We multiply Eq. 2.95 from the left by $X^{(j)T}$ and obtain

$$X^{(j)T}AX^{(i)} = X^{(j)T}\lambda_i X^{(i)} = \lambda_i(X^{(j)T}X^{(i)})$$

The left-hand side is also equal to

$$(AX^{(j)})^T X^{(i)} = (\lambda_j X^{(j)})^T X^{(i)} = \lambda_j(X^{(j)T}X^{(i)})$$

from which follows that

$$(\lambda_i - \lambda_j)X^{(j)T}X^{(i)} = 0$$

Because $\lambda_i \neq \lambda_j$, we conclude that $X^{(j)T}X^{(i)} = 0$, which means orthogonality.

Example 2.19

The matrix $A = \begin{pmatrix} 2 & 1 \\ 1 & 2 \end{pmatrix}$ (Examples 2.17 and 2.18) is symmetric and has the real eigenvalues $\lambda_1 = 1$ and $\lambda_2 = 3$. The eigenvectors $X^{(1)}$ and $X^{(2)}$ are orthogonal:

$$X^{(1)T}X^{(2)} = \frac{1}{\sqrt{2}}(1, -1) \cdot \frac{1}{\sqrt{2}}\begin{pmatrix} 1 \\ 1 \end{pmatrix} = \frac{1}{2}(1 - 1) = 0 \qquad \blacksquare$$

An eigenvalue λ is called p times degenerate, $2 \leqslant p \leqslant n$, if p of the solutions λ_i of Eq. 2.93 are equal to each other. The eigenvectors associated with a degenerate eigenvalue λ are no longer distinct. If A is symmetric, all eigenvectors belonging to λ are elements of a p-dimensional vector space. It is possible to construct p orthonormal eigenvectors belonging to λ by a well-defined procedure (Gram-Schmidt orthogonalization process).[5] These p orthonormal vectors together with the other eigenvectors belonging to nondegenerate eigenvalues form a basis in the n-dimensional vector space.

QUESTIONS

2.1 Try to explain "right" and "left" to a friend over the telephone without relating to objects known to both of you. What are your conclusions?

2.2 State examples which show that the choice of a coordinate system may be suggested by the symmetry properties of the physical system.

[5]See for example, H. Anton, *Elementary Linear Algebra*, Wiley, New York: 3rd ed., 1981; B. Kolman, *Elementary Linear Algebra*, Macmillan, New York: 1970; S. Lang, *Linear Algebra*, Addison-Wesley, Reading, Mass.: 2nd ed., 1971.

PROBLEMS

2.1 Derive the law of cosines for a triangle by applying the inner (scalar, dot) product to the vector sum $\mathbf{a} + \mathbf{b} = \mathbf{c}$.

2.2 Derive the law of sines for a triangle by applying the outer (vector, cross) product to the two vectors $\mathbf{a}$ and $\mathbf{a} + \mathbf{b} = \mathbf{c}$.

2.3 Prove the identities

(a) $$\mathbf{a} \times (\mathbf{b} \times \mathbf{c}) = \mathbf{b}(\mathbf{a} \cdot \mathbf{c}) - \mathbf{c}(\mathbf{a} \cdot \mathbf{b})$$

(b) $$(\mathbf{a} \times \mathbf{b}) \cdot (\mathbf{c} \times \mathbf{d}) = (\mathbf{a} \cdot \mathbf{c})(\mathbf{b} \cdot \mathbf{d}) - (\mathbf{a} \cdot \mathbf{d})(\mathbf{b} \cdot \mathbf{c})$$

2.4 Prove that $\mathbf{a} \cdot (\mathbf{b} \times \mathbf{c})$ is equal to the volume of a parallelepiped with sides a, b, and c.

2.5 Show directly using Eq. 2.3 that the set of basis vectors of the spherical polar coordinate system is orthonormal.

2.6 Assume that f, $\mathbf{g}$, and $\mathbf{h}$ are functions of t. Prove

(a) $$\frac{d}{dt}(f\mathbf{g}) = \frac{df}{dt}\mathbf{g} + f\frac{d\mathbf{g}}{dt}$$

(b) $$\frac{d}{dt}(\mathbf{g} \cdot \mathbf{h}) = \frac{d\mathbf{g}}{dt} \cdot \mathbf{h} + \mathbf{g} \cdot \frac{d\mathbf{h}}{dt}$$

(c) $$\frac{d}{dt}(\mathbf{g} \times \mathbf{h}) = \frac{d\mathbf{g}}{dt} \times \mathbf{h} + \mathbf{g} \times \frac{d\mathbf{h}}{dt}$$

Note that the sequence of the vectors must be maintained when they appear in cross products.

2.7 Let f, $\mathbf{g}$, and h be functions of r. Prove the identities:

(a) $$\nabla \cdot (f\mathbf{g}) = \nabla f \cdot \mathbf{g} + f\nabla \cdot \mathbf{g}$$

(b) $$\nabla f(h) = \frac{df}{dh}\nabla h \quad \text{(chain rule)}$$

(c) $$\nabla \times (f\mathbf{g}) = \nabla f \times \mathbf{g} + f\nabla \times \mathbf{g}$$

2.8 Prove

(a) $$\nabla \times \mathbf{r} = 0$$

(b) $$\nabla r = \mathbf{e}_r \quad (r = |\mathbf{r}|)$$

(c) $$\nabla \frac{1}{r} = -\frac{1}{r^2}\mathbf{e}_r \quad (r \neq 0)$$

2.9 In two dimensions, evaluate (a) $\nabla \ln r \quad (r = |\mathbf{r}|)$ (b) $\nabla^2 \ln r \quad (r \neq 0)$

2.10 In three dimensions, evaluate (a) $\nabla \cdot \mathbf{r}$ (b) $\nabla^2 r$ (c) $\nabla^2 \frac{1}{r}$ $(r \neq 0)$
(d) $\nabla \times (\mathbf{a} \times \mathbf{r})$ $(\mathbf{a} = \text{constant})$

2.11 Evaluate the following integrals:
(a) $\int_C x^2 y^2 \, ds$ where C is the portion of a circle of radius 1 connecting the points $(1, 0)$ and $(0, 1)$.
(b) $\int_C x^2 y^2 \, ds$ where C is the straight line connecting the points $(0, 1)$ and $(1, 0)$.
(c) $\int_C \mathbf{f} \cdot d\mathbf{r}$ where $\mathbf{f} = x\mathbf{i} + y\mathbf{j}$ and c is the straight line from the point $(0, 1)$ to the point $(1, 2)$.

2.12 Prove that the line integral (2.50) is not equal to $\int_{x_1}^{x_2} f_x \, dx + \int_{y_1}^{y_2} f_y \, dy + \int_{z_1}^{z_2} f_z \, dz$ where (x_1, y_1, z_1) and (x_2, y_2, z_2) are the coordinates of the points P_1 and P_2, respectively.

2.13 From the vector functions given below single out the vectors $\mathbf{f}(\mathbf{r})$ for which the line integral is path independent, and find the indefinite line integral

(a) $\qquad\qquad \mathbf{f} = (yz, xz, xy)$

(b) $\qquad\qquad \mathbf{f} = (\sin y, x \cos y - z \sin y, \cos y)$

(c) $\qquad\qquad \mathbf{f} = (\ln z, \ln x, \ln y)$

(d) $\qquad\qquad \mathbf{f} = (e^y, xe^y + e^{-z}, -ye^{-z})$

(e) $\qquad\qquad \mathbf{f} = (e^y, e^z, e^{-x})$

(f) $\qquad\qquad \mathbf{f} = (x^2, \cos y, z \cos y)$

(g) $\qquad\qquad \mathbf{f} = \left(\frac{z}{x}, \frac{z}{y}, \frac{1}{z} \right)$

2.14 For the matrices

$$A = \begin{pmatrix} 1 & 2 & 3 \\ 4 & 5 & 6 \end{pmatrix}, \quad B = \begin{pmatrix} 0 & 1 \\ -1 & 2 \end{pmatrix}, \quad C = \begin{pmatrix} 2 \\ 3 \end{pmatrix}$$

perform the following operations if allowed:

(a) AB (b) BA (c) BC (d) A^{-1} (e) A^T

(f) $A^T A$ (g) AA^T (h) $B^T B$ (i) BB^T

2.15 Show whether or not the product of two symmetric matrices is symmetric.

2.16 Check whether or not the matrix

$$\begin{pmatrix} 0.9 & 0.3\sqrt{2} & -0.1 \\ -0.1 & 0.3\sqrt{2} & 0.9 \\ 0.3\sqrt{2} & -0.8 & 0.3\sqrt{2} \end{pmatrix}$$

is orthogonal.

2.17 Show whether or not the product of two orthogonal matrices is orthogonal.

2.18 Show that the inverse of the unit matrix is the unit matrix.

2.19 Show that square matrices $A \neq 0$ with $A^2 = 0$ have no inverse.

2.20 Assume that A and B are symmetric matrices and C and D are orthogonal. Prove that

(a) $C^{-1}AC$ is symmetric

(b) CD is orthogonal

(c) AB + BA is symmetric

2.21 Prove that the cross product of

(a) two polar vectors is an axial vector

(b) two axial vectors is an axial vector

(c) a polar and an axial vector is a polar vector

2.22 Find the eigenvalues and eigenvectors of the following matrices:

(a) the unit matrix in the three-dimensional space

(b) $B = \begin{pmatrix} 0 & 3 \\ -2 & 5 \end{pmatrix}$ (c) $C = \begin{pmatrix} 4 & -1 & -2 \\ -1 & 1 & -1 \\ -2 & -1 & 4 \end{pmatrix}$ (d) $D = \begin{pmatrix} 1 & 2 & 3 \\ 2 & 2 & 2 \\ 3 & 2 & 1 \end{pmatrix}$

2.23 Show that the diagonal elements of a diagonal matrix are the eigenvalues of the matrix.

2.24 Prove that if λ is an eigenvalue of the matrix A, then λ^2 is an eigenvalue of the matrix A^2.

2.25 Show that the eigenvalues of an orthogonal matrix have unit magnitude: $|\lambda| = 1$ or $\lambda = e^{i\theta}$.

chapter
3 Kinematics of a Particle

*Give me one firm spot on which to stand,
and I will move the earth.*

3.1 INTRODUCTION

Kinematics is the geometrical description of the motion of a physical system in our three-dimensional (flat[1]) space and in time. The system is described in terms of so-called kinematical quantities such as position, displacement, velocity, acceleration, angular velocity, and so forth. Kinematics deals with questions of *how* objects move.

Dynamics is the causal explanation of the motion of a physical system. The quantities needed, which relate to the cause of the motion of the system as we observe it, are called dynamical quantities such as momentum, force, potential energy, and so on. Dynamics, which covers the main part of analytical mechanics, is founded upon a set of basic rules (axioms or postulates) and the knowledge of force laws (or equivalent relationships) from which the general behavior of any mechanical system can be derived. Dynamics deals with the questions of *how* and *why* objects move. It should be clear from Chapter 1 that both kinematics and dynamics require a reference frame.

The description of the motion of a real system like a solid or fluid may turn out to be very complex. In order to facilitate the description one introduces idealizations of the real system.

A particle is an object that is considered to have no extension. The particle's position can be represented by a point in the three-dimensional space. This picture is useful if the size of the object is very small compared with the dimensions of its orbit or surroundings. In dynamics we shall call the point particle a mass point.

A rigid body is an object in which the relative positions of all points remain unchanged at all times. The body is not allowed to undergo deformations.

A system of many discrete particles may sometimes be treated as a continuous mass distribution (continuum). Some physical quantities of the system are

---

[1]Flatness of space is necessary for Newton's mechanics because a curved space-time world is connected with the dynamical theory of gravitation, namely, general relativity.

then introduced as space averages of the corresponding quantities of the discrete system. These averaged quantities are represented by fields. A field is defined as a quantity or a function that depends on space and time coordinates. The function may be a scalar, a vector, a tensor, or any more general mathematical object. Examples of fields are the mass density of the earth's atmosphere (scalar), the gravitational acceleration (vector), and the pressure in a moving viscous fluid (tensor).

Of greatest significance in kinematics is the observation that the general motion of a system can be considered as the superposition of independent special types of motion (see Chapter 9).

A particle (or a specific point of an extended object) in motion traces out a curve (or trajectory). The motion is called rectilinear if the orbit is a straight line, and it is called circular if the orbit is a circle. The combined motion is called translational. These two special types of the motion are sufficient to describe the general motion of a point particle because every infinitesimal section of a curved orbit can be approximated by a section of a circle of a certain radius.

An extended object may rotate about an axis passing through a specific point of the object. In this case of rotational motion, each point of the object undergoes circular motion relative to the axis of rotation. We sometimes refer to the system as a spinning system.

The motion of a system is called oscillatory if the system moves about a state of equilibrium as a result of a perturbation (or displacement) from that state. We talk about wave motion if the system transmits—whether periodically or not—energy and momentum, whereby the system itself does not undergo a permanent change of its state. Possible forms of the motion for some systems are listed in Table 3.1.

Translational motion of a particle and wave motion of a continuum are entirely different forms of the motion. They are mutually exclusive in classical physics. However, the pioneering experiments of J. J. Thomson, Rutherford, Davisson and Germer, and G. P. Thomson have shown that microphysical systems (atoms, electrons) exhibit both particle and continuum properties. Quantum mechanics emerged as a new formalism that allowed physicists to describe the microphysical systems as continua (fields) and particles simultaneously. The existence of the quantum mechanical formalism rests on the fact that the dynamical equations for particle and wave motion possess a common abstract algebraic structure[2] (or symmetry).

3.2 TRANSLATIONAL MOTION OF A PARTICLE

3.2.1 Displacement, Velocity, and Acceleration

The position of a particle is described by the radius vector (position vector) $\mathbf{r}$ which is, in general, a function of the time t: $\mathbf{r} = \mathbf{r}(t)$. Remember that the radius vector $\mathbf{r}$ has a meaning only in a given coordinate system. If the particle moves from a point P to another point Q during the time interval Δt, its position vector changes by

$$\Delta\mathbf{r} = \mathbf{r}(t + \Delta t) - \mathbf{r}(t) \tag{3.1}$$

[2]See Appendix 4.

Table 3.1 Possible Forms of the Motion for Some Systems

Form of the Motion	Mass Point	Rigid Body	Real Solid
Rectilinear motion	X	X	X
Circular motion	X	X	X
Rotation		X	X
Oscillation			X

In contrast to **r**, the displacement vector $\Delta\mathbf{r}$ is independent of the chosen coordinate system (see Figure 3.1) because we have

$$\Delta\mathbf{r} = \mathbf{r}(t + \Delta t) - \mathbf{r}(t)$$

$$= \mathbf{r}'(t + \Delta t) - \mathbf{r}'(t) \tag{3.2}$$

where $\mathbf{r}'(t)$ is the position vector of the point P in S'.

The velocity and acceleration vectors are defined as the first and second derivatives of the position vector **r** with respect to the time t:

$$\mathbf{v} = \frac{d\mathbf{r}}{dt}, \quad \mathbf{a} = \frac{d\mathbf{v}}{dt} = \frac{d^2\mathbf{r}}{dt^2}$$

Because the definition involves the (infinitesimal) displacement $d\mathbf{r}$, both **v** and **a** are independent of the coordinate system. It is customary and convenient to denote a time derivative by a dot over the letter representing the physical quantity. Then

$$\mathbf{v} = \frac{d\mathbf{r}}{dt} = \dot{\mathbf{r}} \tag{3.3}$$

$$\mathbf{a} = \frac{d\mathbf{v}}{dt} = \dot{\mathbf{v}} = \ddot{\mathbf{r}} \tag{3.4}$$

The magnitude of **r**, $r = |\mathbf{r}|$, is the distance between the origin of the coordinate system and the position of the particle. The magnitude of **v**, $v = |\mathbf{v}|$ is the speed of the particle. No special name has been given to the magnitude of **a**.

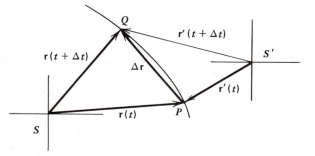

Figure 3.1 Displacement vector.

Note that

$$v = \left| \frac{d\mathbf{r}}{dt} \right| \neq \frac{d|\mathbf{r}|}{dt} = \frac{dr}{dt} = \dot{r}$$

Two identities are very useful:

$$r\dot{r} = \frac{1}{2}\frac{d}{dt}(r^2) = \frac{1}{2}\frac{d}{dt}(\mathbf{r} \cdot \mathbf{r}) = \mathbf{r} \cdot \dot{\mathbf{r}} \qquad (3.5)$$

$$v\dot{v} = \frac{1}{2}\frac{d}{dt}(v^2) = \frac{1}{2}\frac{d}{dt}(\mathbf{v} \cdot \mathbf{v}) = \mathbf{v} \cdot \dot{\mathbf{v}} \qquad (3.6)$$

The product $\mathbf{r} \cdot \dot{\mathbf{r}}$ is equal to zero in the following cases:

(a) $\mathbf{r} = 0$. The particle is at the origin (trivial).
(b) $\dot{\mathbf{r}} = 0$. The vector $\mathbf{r}$ is constant, and the particle is at rest.
(c) $\mathbf{r}$ is orthogonal to $\dot{\mathbf{r}}$ and hence also orthogonal to $d\mathbf{r}$ ($\mathbf{r} \perp \dot{\mathbf{r}}, \mathbf{r} \perp d\mathbf{r}$). Then the magnitude r is a constant, that is, $\dot{r} = 0$, and the particle undergoes circular motion.

Equivalent conclusions can be drawn for $\mathbf{v}$ and $\dot{\mathbf{v}}$ from the identity (3.6).

3.2.2 Velocity and Acceleration in Special Coordinate Systems
We now derive the coordinates of the velocity and acceleration vectors in special coordinate systems.

Cartesian Coordinate System. The right-handed triad of basis vectors is $\{\mathbf{i}, \mathbf{j}, \mathbf{k}\}$. In a nonrotating and nonaccelerating Cartesian coordinate system the basis vectors are constant, and we obtain

$$\mathbf{r} = (x, y, z) \qquad (3.7a)$$

$$\mathbf{v} = (\dot{x}, \dot{y}, \dot{z}) \qquad (3.7b)$$

$$v^2 = \dot{x}^2 + \dot{y}^2 + \dot{z}^2 \qquad (3.7c)$$

$$\mathbf{a} = (\ddot{x}, \ddot{y}, \ddot{z}) \qquad (3.7d)$$

Example 3.1

Let
$$\mathbf{r}(t) = 2t^3\mathbf{i} - 3t\mathbf{j} + 4\mathbf{k}.$$

Then
$$\mathbf{v}(t) = 6t^2\mathbf{i} - 3\mathbf{j}$$

$$v^2 = 36t^4 + 9$$

$$\mathbf{a}(t) = 12t\mathbf{i} \qquad \blacksquare$$

Circular Cylindrical Coordinate System. The right-handed triad of basis vectors is $\{e_\rho, e_\varphi, k\}$. The basis vectors e_ρ and e_φ are not constant. Applying Eq. 2.31 to r and then once more to v, we obtain

$$\mathbf{r} = (\rho, 0, z) \tag{3.8a}$$

$$\mathbf{v} = (\dot\rho, \rho\dot\varphi, \dot z) \tag{3.8b}$$

$$v^2 = \dot\rho^2 + \rho^2\dot\varphi^2 + \dot z^2 \tag{3.8c}$$

$$\mathbf{a} = (\ddot\rho - \rho\dot\varphi^2, 2\dot\rho\dot\varphi + \rho\ddot\varphi, \ddot z) \tag{3.8d}$$

Plane Polar Coordinate System. The expressions for $\mathbf{r}$, $\mathbf{v}$, and $\mathbf{a}$ are easily obtained from Eqs. 8a through 8d by setting $z = 0$, $\dot z = 0$, and $\ddot z = 0$.

$$\mathbf{r} = (\rho, 0) \tag{3.9a}$$

$$\mathbf{v} = (\dot\rho, \rho\dot\varphi) \tag{3.9b}$$

$$v^2 = \dot\rho^2 + \rho^2\dot\varphi^2 \tag{3.9c}$$

$$\mathbf{a} = (\ddot\rho - \rho\dot\varphi^2, 2\dot\rho\dot\varphi + \rho\ddot\varphi) \tag{3.9d}$$

Spherical Polar Coordinate System. The right-handed triad of basis vectors is $\{e_r, e_\theta, e_\varphi\}$. The basis vectors are not constant. Applying Eq. 2.36 to r and then once more to v we obtain

$$\mathbf{r} = (r, 0, 0) \tag{3.10a}$$

$$\mathbf{v} = (\dot r, r\dot\theta, r\sin\theta\dot\varphi) \tag{3.10b}$$

$$v^2 = \dot r^2 + r^2(\dot\theta^2 + \sin^2\theta\dot\varphi^2) \tag{3.10c}$$

$$\mathbf{a} = (\ddot r - r(\dot\theta^2 + \sin^2\theta\dot\varphi^2), 2\dot r\dot\theta + r(\ddot\theta - \sin\theta\cos\theta\dot\varphi^2),$$

$$2\dot r\sin\theta\dot\varphi + r(2\cos\theta\dot\theta\dot\varphi + \sin\theta\ddot\varphi)) \tag{3.10d}$$

Example 3.2

Let $r = At$, $\theta = \omega t$, $\varphi = \varphi_0$, where A, ω, and φ_0 are constants. Then

$$\mathbf{r} = (At, 0, 0)$$

$$\mathbf{v} = (A, A\omega t, 0)$$

$$v^2 = A^2(1 + \omega^2 t^2)$$

$$\mathbf{a} = (-At\omega^2, 2A\omega, 0) \qquad \blacksquare$$

3.2.3 Path Coordinates

A right-handed orthonormal basis can be attached in a unique way to each point P of a general smooth curve (the trajectory of a moving particle) in the three-dimensional space. If $ds = |d\mathbf{r}|$ is the magnitude of the infinitesimal displacement vector $d\mathbf{r}$, then the vector

$$\boldsymbol{\tau} = \frac{d\mathbf{r}}{ds} \tag{3.11}$$

is a unit vector in the direction of the infinitesimal displacement:

$$\boldsymbol{\tau} \cdot \boldsymbol{\tau} = \frac{d\mathbf{r}}{ds} \cdot \frac{d\mathbf{r}}{ds} = \frac{ds}{ds}\frac{ds}{ds}\cos 0 = 1$$

$\boldsymbol{\tau}$ is called the tangent unit vector. Both $\boldsymbol{\tau}$ and an infinitesimal portion of the curve through P lie within a plane called the osculating[3] plane. (There are infinitely many osculating planes if the curve through P is a section of a straight line.) A second unit vector $\mathbf{n}$ called the normal vector is defined by the conditions that $\mathbf{n}$ lies in the osculating plane, $\mathbf{n}$ is orthogonal to $\boldsymbol{\tau}$, and $\mathbf{n}$ is directed toward the concave side of the trajectory. A third unit vector $\mathbf{b}$, the binormal vector is defined by

$$\mathbf{b} = \boldsymbol{\tau} \times \mathbf{n}$$

Then $\boldsymbol{\tau}$, $\mathbf{n}$, and $\mathbf{b}$ form a right-handed orthonormal basis of a coordinate system called the path coordinate system. Note that the path coordinate system is fixed to the point P on the trajectory of the particle and not to the particle itself.

Equation 3.11 can be written in the form

$$d\mathbf{r} = ds\,\boldsymbol{\tau}$$

from which follows

$$\mathbf{v} = \dot{\mathbf{r}} = \dot{s}\boldsymbol{\tau} = v\boldsymbol{\tau} = (v, 0, 0) \tag{3.12}$$

where $v = \dot{s}$ is the instantaneous speed of the point particle.

If the particle undergoes the displacement $d\mathbf{r}$, the vector $\boldsymbol{\tau}$ is changed by $d\boldsymbol{\tau}$. The infinitesimal vector $d\boldsymbol{\tau}$ lies in the osculating plane. Because $\boldsymbol{\tau}$ is also a unit vector, then, by the conclusion stated following Eq. 2.25, $d\boldsymbol{\tau}$ must be orthogonal to $\boldsymbol{\tau}$, that is, parallel to $\mathbf{n}$:

$$d\boldsymbol{\tau} = d\varepsilon\,\mathbf{n}, \tag{3.13}$$

where $d\varepsilon$ is the infinitesimal angle (measured in radians) between $\boldsymbol{\tau}$ and $\boldsymbol{\tau} + d\boldsymbol{\tau}$ (see Figures 3.2a and 3.2b. The time rate of change of $\boldsymbol{\tau}$ is then

$$\dot{\boldsymbol{\tau}} = \frac{d\boldsymbol{\tau}}{dt} = \frac{d\varepsilon}{dt}\mathbf{n} = \frac{d\varepsilon}{ds}\frac{ds}{dt}\mathbf{n} = \left(\frac{ds}{d\varepsilon}\right)^{-1} v\mathbf{n} \tag{3.14}$$

---

[3] *Osculum* (Latin) = kiss.

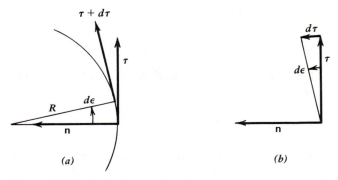

Figure 3.2 Tangent and normal vectors.

The differential quotient $ds/d\varepsilon$ is equal to the radius R of the circle osculating the curve at the point P. (Remember that the angle is defined by the arc length of a circle of radius R divided by R). R is called the radius of curvature of the curve at P. Equation 3.14 becomes

$$\dot{\tau} = \frac{v}{R}\mathbf{n} \tag{3.15}$$

Using (3.12) and (3.15) we obtain for the acceleration the result

$$\mathbf{a} = \dot{\mathbf{v}} = \dot{v}\tau + v\dot{\tau}$$

$$= \dot{v}\tau + \frac{v^2}{R}\mathbf{n} = \left(\dot{v}, \frac{v^2}{R}, 0\right) \tag{3.16}$$

The component $\dot{v}\tau$, the tangential acceleration, is the rate of change of the speed in the direction of the motion. The other term $(v^2/R)\mathbf{n}$, the centripetal acceleration, is the rate of change of the direction of the velocity. It is directed toward the center of the osculating circle. The magnitude of $\mathbf{a}$ is given by

$$a = \sqrt{\dot{v}^2 + \frac{v^4}{R^2}}$$

3.3   CIRCULAR MOTION OF A PARTICLE

A rotation of a point P is uniquely described by an axis of rotation, the rotation angle θ, and the sense of rotation (see Figure 3.3). Rotations are expressed mathematically by orthogonal matrices with determinant $+1$. We denote by $\mathbf{R}$ a rotation matrix that describes a coordinate transformation and by $\mathbf{R}^T = \mathbf{R}^{-1}$ the associated rotation matrix that describes a point transformation (see Section 2.9). For coordinate transformations the coordinates $\mathbf{X}$ of P (in matrix form) relative to the coordinate system S are transformed into the coordinates $\mathbf{X}'$ of the same point relative to the rotated coordinate system S' by

$$X' = RX$$

For point transformations the position vector X of P is transformed into the position vector X' of the point P' by

$$X' = R^T X \qquad (3.17)$$

The rotation matrices (for coordinate transformations) about the Cartesian coordinate axes have the form

$$R_x = \begin{pmatrix} 1 & 0 & 0 \\ 0 & \cos\theta_x & \sin\theta_x \\ 0 & -\sin\theta_x & \cos\theta_x \end{pmatrix},$$

$$R_y = \begin{pmatrix} \cos\theta_y & 0 & -\sin\theta_y \\ 0 & 1 & 0 \\ \sin\theta_y & 0 & \cos\theta_y \end{pmatrix},$$

$$R_z = \begin{pmatrix} \cos\theta_z & \sin\theta_z & 0 \\ -\sin\theta_z & \cos\theta_z & 0 \\ 0 & 0 & 1 \end{pmatrix} \qquad (3.18)$$

where θ_x, θ_y, and θ_z are the corresponding rotation angles. The matrix R_z is identical with the matrix (2.83) because $\theta_z = \varphi$. The matrices R_x and R_y are obtained from R_z by cyclic permutations of the indices:

$$R_z \to R_x \quad \text{if} \quad \theta_z \to \theta_x \colon 3 \to 1, 1 \to 2, 2 \to 3$$

$$R_z \to R_y \quad \text{if} \quad \theta_z \to \theta_y \colon 3 \to 2, 2 \to 1, 1 \to 3$$

(For example, $R_{z_{21}} = -\sin\theta_z \to R_{y_{13}} = -\sin\theta_y$.) The corresponding matrices

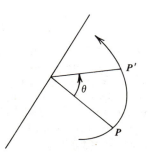

Figure 3.3 Circular motion.

52   KINEMATICS OF A PARTICLE

describing point transformations are

$$\mathbf{R}_x^T = \begin{pmatrix} 1 & 0 & 0 \\ 0 & \cos\theta_x & -\sin\theta_x \\ 0 & \sin\theta_x & \cos\theta_x \end{pmatrix},$$

$$\mathbf{R}_y^T = \begin{pmatrix} \cos\theta_y & 0 & \sin\theta_y \\ 0 & 1 & 0 \\ -\sin\theta_y & 0 & \cos\theta_y \end{pmatrix},$$

$$\mathbf{R}_z^T = \begin{pmatrix} \cos\theta_z & -\sin\theta_z & 0 \\ \sin\theta_z & \cos\theta_z & 0 \\ 0 & 0 & 1 \end{pmatrix} \tag{3.19}$$

Let $\mathbf{R}_1$ and $\mathbf{R}_2$ denote two arbitrary *finite* rotations. Because matrix multiplication is not commutative, $\mathbf{R}_1\mathbf{R}_2 \neq \mathbf{R}_2\mathbf{R}_1$, the rotation $\mathbf{R}_1$ followed by $\mathbf{R}_2$, and the rotation $\mathbf{R}_2$ followed by $\mathbf{R}_1$ are different.

Example 3.3

Let $\theta_x = \theta_y = 90°$. For point transformations we find

$$\mathbf{R}_y^T\mathbf{R}_x^T = \begin{pmatrix} 0 & 0 & 1 \\ 0 & 1 & 0 \\ -1 & 0 & 0 \end{pmatrix}\begin{pmatrix} 1 & 0 & 0 \\ 0 & 0 & -1 \\ 0 & 1 & 0 \end{pmatrix} = \begin{pmatrix} 0 & 1 & 0 \\ 0 & 0 & -1 \\ -1 & 0 & 0 \end{pmatrix}$$

whereas

$$\mathbf{R}_x^T\mathbf{R}_y^T = \begin{pmatrix} 1 & 0 & 0 \\ 0 & 0 & -1 \\ 0 & 1 & 0 \end{pmatrix}\begin{pmatrix} 0 & 0 & 1 \\ 0 & 1 & 0 \\ -1 & 0 & 0 \end{pmatrix} = \begin{pmatrix} 0 & 0 & 1 \\ 1 & 0 & 0 \\ 0 & 1 & 0 \end{pmatrix}$$

The point $(0,0,1)$ is transformed by $\mathbf{R}_y^T\mathbf{R}_x^T$ into the point $(0,-1,0)$ and by $\mathbf{R}_x^T\mathbf{R}_y^T$ into the point $(1,0,0)$ (see Figure 3.4). ∎

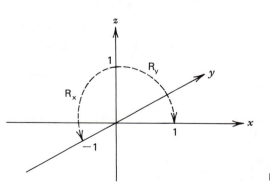

Figure 3.4 Example 3.3.

The mathematical treatment of *infinitesimal* rotations is much simpler. Because an infinitesimal rotation differs from no rotation, described by the unit matrix 1, by an infinitesimal term, we may write

$$R = 1 + \theta S \qquad (3.20)$$

where θ is the infinitesimal rotation angle and S is a matrix describing the rotation axis and the sense of rotation. Mathematically, the right side of Eq. 3.20 consists of the first two terms of the McLaurin expansion[4] of the matrix R about the angle $\theta = 0$. Therefore, S is related to R simply by

$$S = \left. \frac{dR}{d\theta} \right|_{\theta=0} \qquad (3.21)$$

Applying this equation to the matrices R_x, R_y, R_z, Eq. 3.18, we obtain

$$S_x = \left. \frac{dR_x(\theta_x)}{d\theta_x} \right|_{\theta_x=0} = \left. \begin{pmatrix} 0 & 0 & 0 \\ 0 & -\sin\theta_x & \cos\theta_x \\ 0 & -\cos\theta_x & -\sin\theta_x \end{pmatrix} \right|_{\theta_x=0} = \begin{pmatrix} 0 & 0 & 0 \\ 0 & 0 & 1 \\ 0 & -1 & 0 \end{pmatrix}, \qquad (3.22)$$

$$S_y = \begin{pmatrix} 0 & 0 & -1 \\ 0 & 0 & 0 \\ 1 & 0 & 0 \end{pmatrix} \qquad (3.23)$$

and

$$S_z = \begin{pmatrix} 0 & 1 & 0 \\ -1 & 0 & 0 \\ 0 & 0 & 0 \end{pmatrix} \qquad (3.24)$$

Note that these matrices are antisymmetric.

Let us now perform two infinitesimal rotations in succession. Comparing

$$R_2 R_1 = (1 + \theta_2 S_2)(1 + \theta_1 S_1) = 1 + \theta_1 S_1 + \theta_2 S_2 + \theta_1 \theta_2 S_2 S_1$$

with

$$R_1 R_2 = (1 + \theta_1 S_1)(1 + \theta_2 S_2) = 1 + \theta_2 S_2 + \theta_1 S_1 + \theta_1 \theta_2 S_1 S_2$$

we see that the two matrix products are equal to each other if we neglect their last terms, which are small. We have found the important result that two infinitesimal rotations commute,

$$R_1 R_2 = R_2 R_1$$

---

[4]The McLaurin expansion of a matrix valued function $M(x)$ is no more difficult than the McLaurin expansion of an ordinary function:

$$M(x) = M(0) + \frac{1}{1!} \left. \frac{dM(x)}{dx} \right|_{x=0} x + \frac{1}{2!} \left. \frac{d^2 M(x)}{dx^2} \right|_{x=0} x^2 + \cdots$$

and that they may therefore be performed simultaneously. Consequently, any infinitesimal rotation may be written in the form

$$R(\theta) = 1 + \theta_x S_x + \theta_y S_y + \theta_z S_z \tag{3.25}$$

Because the matrices S_i are antisymmetric, we also have

$$R^T(\theta) = 1 - (\theta_x S_x + \theta_y S_y + \theta_z S_z) \tag{3.26}$$

Trivially, $R(0) = R^T(0) = 1$.

We wish to find the velocity of a point P that undergoes a rotation. Because the point P moves, we are dealing here with a point transformation expressed by Eq. 3.17. We generalize our problem slightly by asking for the time rate of change of any time dependent vector $Q(t)$ (in matrix form) which is simultaneously subjected to a rotation $R^T(\theta)$. The time rate of change of $Q(t)$ is then given by

$$\frac{dQ(t)}{dt} = \lim_{\Delta t \to 0} \frac{1}{\Delta t} \left[R^T(\Delta\theta)Q(t + \Delta t) - R^T(0)Q(t) \right]$$

$$= \lim_{\Delta t \to 0} \frac{1}{\Delta t} \left[(R^T(\Delta\theta) - R^T(0))Q(t + \Delta t) + R^T(0)(Q(t + \Delta t) - Q(t)) \right]$$

$$= \lim_{\Delta t \to 0} \frac{1}{\Delta t} (R^T(\Delta\theta) - 1)Q(t) + \dot{Q}(t) \tag{3.27}$$

Let us introduce the matrix of angular velocity Ω by

$$\Omega = \lim_{\Delta t \to 0} \frac{1}{\Delta t} (R^T(\Delta\theta) - 1)$$

$$= \frac{dR^T}{dt}\bigg|_{\theta=0}$$

$$= - (\dot{\theta}_x S_x + \dot{\theta}_y S_y + \dot{\theta}_z S_z) \tag{3.28}$$

Defining the angular velocities by

$$\omega_x = \dot{\theta}_x, \quad \omega_y = \dot{\theta}_y, \quad \omega_z = \dot{\theta}_z$$

and using Eqs. 3.22, 3.23, 3.24, and 3.28, we obtain Ω in the explicit form

$$\Omega = \begin{pmatrix} 0 & -\omega_z & \omega_y \\ \omega_z & 0 & -\omega_x \\ -\omega_y & \omega_x & 0 \end{pmatrix} \tag{3.29}$$

which shows that Ω is an antisymmetric matrix. Equation 3.27 may be written

$$\frac{dQ}{dt} = \Omega Q + \dot{Q} \tag{3.30}$$

The total rate of change dQ/dt consists of two parts: ΩQ is the rate of change of Q due to rotation, and $\dot{Q}$ is the time rate of change of Q if the rotation were absent.

In order to find the coordinates of dQ/dt, we apply (3.29) to (3.30):

$$\frac{d\mathbf{Q}}{dt} = \begin{pmatrix} \dfrac{dq_x}{dt} \\ \dfrac{dq_y}{dt} \\ \dfrac{dq_z}{dt} \end{pmatrix} = \begin{pmatrix} 0 & -\omega_z & \omega_y \\ \omega_z & 0 & -\omega_x \\ -\omega_y & \omega_x & 0 \end{pmatrix} \begin{pmatrix} q_x \\ q_y \\ q_z \end{pmatrix} + \begin{pmatrix} \dot{q}_x \\ \dot{q}_y \\ \dot{q}_z \end{pmatrix}$$

$$= \begin{pmatrix} \omega_y q_z - \omega_z q_y \\ \omega_z q_x - \omega_x q_z \\ \omega_x q_y - \omega_y q_x \end{pmatrix} + \begin{pmatrix} \dot{q}_x \\ \dot{q}_y \\ \dot{q}_z \end{pmatrix} \tag{3.31}$$

Combining the angular velocities to a three-component quantity

$$\boldsymbol{\omega} = (\omega_x, \omega_y, \omega_z)$$

we see that Eq. 3.31 is equivalent to

$$\frac{d\mathbf{q}}{dt} = \boldsymbol{\omega} \times \mathbf{q} + \dot{\mathbf{q}} \tag{3.32}$$

Because of the one-to-one correspondence between the components of Ω and $\boldsymbol{\omega}$, the quantity $\boldsymbol{\omega}$ must be an axial vector (see Section 2.11). Also, because $\boldsymbol{\omega}$ is defined by the infinitesimal rotations, the addition of two angular velocity vectors must be commutative, $\boldsymbol{\omega}_1 + \boldsymbol{\omega}_2 = \boldsymbol{\omega}_2 + \boldsymbol{\omega}_1$, which is in agreement with the general vector rules.

An important consequence of (3.32) is the equation

$$\frac{d\boldsymbol{\omega}}{dt} = \dot{\boldsymbol{\omega}} \tag{3.33}$$

which is needed in rigid body mechanics (see Section 9.2).

The one-to-one correspondence between the components of Ω and $\boldsymbol{\omega}$ allows us to describe rotational kinematics in two different but equivalent ways. Although the matrix formulation seems to be more natural, the vector formalism is mathematically simpler.

Equation 3.32 may be derived in the vector formalism as follows: The basis vectors $\mathbf{e}_i$ are unit vectors and can change only their directions. Because for any set of orthonormal basis vectors $\mathbf{e}_i$ holds

$$0 = \frac{d\delta_{ij}}{dt} = \frac{d\mathbf{e}_i \cdot \mathbf{e}_j}{dt} = \mathbf{e}_i \cdot \frac{d\mathbf{e}_j}{dt} + \mathbf{e}_j \cdot \frac{d\mathbf{e}_i}{dt} \tag{3.34}$$

the most general form of the time derivative is

$$\frac{d\mathbf{e}_i}{dt} = \boldsymbol{\omega} \times \mathbf{e}_i$$

with the *same* vector $\boldsymbol{\omega}$ for all $\mathbf{e}_i$. Otherwise, the right side of (3.34) would not be equal to zero. The vector $\boldsymbol{\omega}$, which is an axial vector and describes the speed and direction of the rotation of the basis vectors, must therefore be the angular velocity vector. Writing

$$\mathbf{q}(t) = \sum_{j=1}^{3} q_j(t)\mathbf{e}_j(t)$$

for any vector $\mathbf{q}(t)$ and applying the product rule, we obtain

$$\frac{d\mathbf{q}}{dt} = \sum_{j=1}^{3} q_j \boldsymbol{\omega} \times \mathbf{e}_j + \sum_{j=1}^{3} \dot{q}_j \mathbf{e}_j$$

$$= \boldsymbol{\omega} \times \mathbf{q} + \dot{\mathbf{q}}$$

which is Eq. 3.32.

The second derivative of $\mathbf{Q}$ is obtained from (3.28) if $\mathbf{Q}$ is replaced by $d\mathbf{Q}/dt$:

$$\frac{d^2\mathbf{Q}}{dt^2} = \Omega \frac{d\mathbf{Q}}{dt} + \left(\frac{d\mathbf{Q}}{dt}\right)^{\cdot}$$

$$= \Omega(\Omega\mathbf{Q} + \dot{\mathbf{Q}}) + (\Omega\mathbf{Q} + \dot{\mathbf{Q}})^{\cdot}$$

$$= (\Omega^2 + \dot{\Omega})\mathbf{Q} + 2\Omega\dot{\mathbf{Q}} + \ddot{\mathbf{Q}} \tag{3.35}$$

This equation reads in the vector formalism

$$\frac{d^2\mathbf{q}}{dt^2} = \boldsymbol{\omega} \times (\boldsymbol{\omega} \times \mathbf{q}) + \dot{\boldsymbol{\omega}} \times \mathbf{q} + 2\boldsymbol{\omega} \times \dot{\mathbf{q}} + \ddot{\mathbf{q}}, \tag{3.36}$$

which follows directly from (3.32), when $\mathbf{q}$ is there replaced by $d\mathbf{q}/dt$. We repeat that (3.30) and (3.35), or equivalently (3.32) and (3.36), hold for any vector $\mathbf{Q}$ (or $\mathbf{q}$). These equations are used for deriving the equations of motion in rotating (noninertial) reference frames.

Returning to the original problem of finding the velocity of a point P that undergoes a rotation only, we note that the position vector $\mathbf{r}$ does not *explicitly* depend on the time. It then follows from Eqs. 3.34 and 3.36 that the velocity and acceleration vectors are given by

$$\frac{d\mathbf{r}}{dt} = \boldsymbol{\omega} \times \mathbf{r} \tag{3.37}$$

and

$$\frac{d^2\mathbf{r}}{dt^2} = \boldsymbol{\omega} \times (\boldsymbol{\omega} \times \mathbf{r}) + \dot{\boldsymbol{\omega}} \times \mathbf{r}, \tag{3.38}$$

respectively. The relationship between $\boldsymbol{\omega}$, $\mathbf{r}$, and $\mathbf{v}$ is shown in Figure 3.5. Because for circular motion (fixed axis) the vectors $\boldsymbol{\omega}$ and $\dot{\boldsymbol{\omega}}$ have the same (or opposite) direction, the term $\dot{\boldsymbol{\omega}} \times \mathbf{r}$ represents the tangential acceleration of the particle. And because by (3.37) $\boldsymbol{\omega}$ is orthogonal to $\mathbf{v}$, the term $\boldsymbol{\omega} \times (\boldsymbol{\omega} \times \mathbf{r})$ is the centripetal acceleration (see also Eq. 3.16).

3.4 RELATIVE MOTION

We consider two particles described by the position vectors $\mathbf{r}_1$ and $\mathbf{r}_2$ and the velocities $\dot{\mathbf{r}}_1$ and $\dot{\mathbf{r}}_2$, respectively (see Figures 3.6 and 3.7). The relative position vector $\mathbf{r}_{12}$ is defined by

$$\mathbf{r}_{12} = \mathbf{r}_2 - \mathbf{r}_1 \tag{3.39}$$

and the relative velocity by

$$\mathbf{v}_{12} = \dot{\mathbf{r}}_{12} = \dot{\mathbf{r}}_2 - \dot{\mathbf{r}}_1 \tag{3.40}$$

The vector $\mathbf{r}_{12}$ is the position vector of particle 2 described in a coordinate system attached to particle 1. Similarly, $\mathbf{v}_{12}$ is the velocity of the second particle relative to the first.

We now identify P_1 with the origin O' of a coordinate system S' and consider P_2 as some point P in space (see Figure 3.8). Then we have

$$\mathbf{R} = \mathbf{r}_1, \quad \mathbf{r} = \mathbf{r}_2, \quad \mathbf{r}' = \mathbf{r}_{12},$$

and Eqs. 3.39 and 3.40 become

$$\mathbf{r}' = \mathbf{r} - \mathbf{R} \tag{3.41}$$

$$\dot{\mathbf{r}}' = \dot{\mathbf{r}} - \dot{\mathbf{R}} \tag{3.42}$$

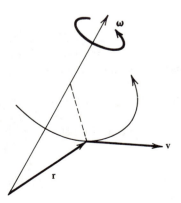

Figure 3.5 Angular velocity vector.

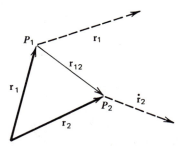

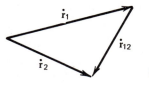

Figure 3.6 Relative position vector.

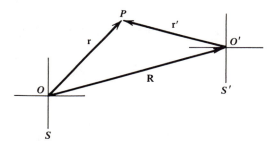

Figure 3.7 Relative velocity vector.

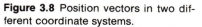

Figure 3.8 Position vectors in two different coordinate systems.

or[5] inversely,

$$\mathbf{r} = \mathbf{r}' + \mathbf{R}$$
$$\dot{\mathbf{r}} = \dot{\mathbf{r}}' + \dot{\mathbf{R}}$$

Example 3.4

A wheel of radius A rolls along a flat surface at constant speed (see Figure 3.9). What are the position and velocity vectors of some point on the rim of the wheel relative to the surface? The wheel is the moving system. The origin O' of the moving coordinate system S' may be placed at the center of the wheel. A point on the rim of the wheel is described by the radius vector

$$\mathbf{r}' = A \cos \omega t \mathbf{i} + A \sin \omega t \mathbf{j}$$

The relative position vector $\mathbf{R}$ from the origin O of the coordinate system S

[5]The proper identification of the three position vectors and velocities in practical examples seems difficult to some students. If these difficulties are not resolved in classical mechanics, students will encounter more problems in special relativity where improper identification of the coordinates can lead to paradoxical results.

located at the surface to the point O' is given by

$$\mathbf{R} = vt\mathbf{i} + A\mathbf{j} = A\omega t\mathbf{i} + A\mathbf{j}$$

Hence,

$$\mathbf{r} = \mathbf{r}' + \mathbf{R} = A(\cos \omega t + \omega t)\mathbf{i} + A(\sin \omega t + 1)\mathbf{j}$$

and

$$\mathbf{v} = \dot{\mathbf{r}} = A\omega(-\sin \omega t + 1)\mathbf{i} + A\omega \cos \omega t\mathbf{j} \qquad \blacksquare$$

If the velocity $\dot{\mathbf{R}}$ is a constant, $\dot{\mathbf{R}} = \mathbf{V}$, then (3.41) can be written

$$\mathbf{r}' = \mathbf{r} - \mathbf{V}t \qquad (3.43)$$

It is assumed that the time coordinates t and t' measured in S and S', respectively, are the same, that is, that the time is universal:

$$t = t' \qquad (3.44)$$

Equations 3.43 and 3.44 constitute the Galilei transformations. If $\mathbf{V} = (V, 0, 0)$, $V = $ const., they reduce to

$$x' = x - Vt$$
$$y' = y$$
$$z' = z$$
$$t' = t \qquad (3.45)$$

and

$$v'_x = v_x - V$$
$$v'_y = v_y$$
$$v'_z = v_z \qquad (3.46)$$

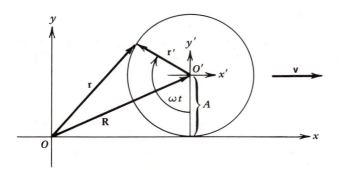

Figure 3.9 Example 3.4.

Until the end of the last century it was generally believed that the Galilei transformations were the correct transformations between any two reference frames moving at constant velocity **V** relative to each other (inertial frames, see Section 4.2). According to Einstein, however, Eqs. 3.46 are not applicable to the speed of light, and $c' = c - V$ had to be replaced by

$$c' = c \tag{3.47}$$

This condition made it necessary to abandon the Galilei transformation and to search for a more general transformation that also involves the time coordinate. This new set of equations corresponding to (3.45) was found to be

$$x' = \gamma(V)(x - Vt)$$
$$y' = y$$
$$z' = z$$
$$t' = \gamma(V)\left(t - \frac{V}{c^2}x\right) \tag{3.48}$$

where

$$\gamma(V) = \frac{1}{\sqrt{1 - \left(\frac{V}{c}\right)^2}}$$

Equations 3.48 represent a special case of the so-called Lorentz transformations[6] which form the general transformation group of special relativity.[7] The fact that the Lorentz transformation (3.48) goes over into the Galilei transformation (3.45) for velocities $V \ll c$, is sometimes referred to as the correspondence principle of special relativity. The development of a new theory that generalizes an already existing theory is always guided by a correspondence principle.

PROBLEMS

3.1   Given below are the coordinates of position vectors as functions of the time. Find the velocity and acceleration vectors in the given coordinate systems. Also find the speed of the particles. A, B, C, and ω are constants.

(a) $x = A \cos \omega t$ $y = A \sin \omega t$ $z = Bt$
(b) $\rho = At$ $\varphi = B$ $z = Ct^2$
(c) $r = A$ $\theta = \omega t$ $\varphi = \frac{1}{2}\omega t$

---

[6]The Lorentz transformations were derived independently by W. Voigt (1887), H. A. Lorentz (1904), A. Einstein (1905), and H. Poincare (1905).
[7]Some texts on relativity are A. P. French, *Special Relativity*, W. W. Norton and Co., New York, 1968; E. F. Taylor, J. A. Wheeler, *Space-Time Physics*, W. H. Freeman & Co., San Francisco, 1966; M. Born, *Einstein's Theory of Relativity*, Dover, New York, 1962; W. Rindler, *Essential Relativity*, Van Nostrand-Reinhold Co., New York, 1969.

3.2 Find the coordinates of $\ddot{\mathbf{r}}$ in

    (a) circular cylindrical coordinates
    (b) spherical polar coordinates

3.3 Show that the tangential component of the acceleration can be written in the form

$$a_t = \frac{\mathbf{a} \cdot \mathbf{v}}{v}$$

3.4 How, if at all possible, could you express the position vector in the path coordinate system?

3.5 A particle oscillates on a circle of radius R. The path coordinate is given by $s = R \sin \omega t$. For which values of t is the magnitude of the acceleration largest? Smallest? What are the values of the speed and the tangential component of the acceleration at those times?

3.6 A particle moves on a circle of radius R following the equation

$$s(t) = R \ln\left(\frac{v_0 t}{R} + 1\right)$$

    (a) Find the speed of the particle.
    (b) Find the magnitude of the acceleration of the particle.

3.7 A particle moves on a cycloid

$$x(t) = A(bt - \sin bt)$$

$$y(t) = A(1 - \cos bt)$$

    (a) Find $a_t(t)$ and $a_n(t)$.
    (b) Find the points on the orbit where a_n has a maximum.
    (c) What are the values of a_t and a_n at $t = 0$?

3.8 Show that $(\Omega^2)_{ij} = \omega_i \omega_j - \omega^2 \delta_{ij}$ where $\omega^2 = \boldsymbol{\omega} \cdot \boldsymbol{\omega}$.

3.9 Derive Eq. 3.36 in the vector formalism.

3.10 Calculate $d^2\boldsymbol{\omega}/dt^2$ from Eq. 3.36.

3.11 The radius vectors of two particles are given by

$$\mathbf{r}_1 = A(\cos \omega t \mathbf{i} + \sin \omega t \mathbf{j})$$

$$\mathbf{r}_2 = A(\cos \tfrac{1}{2}\omega t \mathbf{i} + \sin \tfrac{1}{2}\omega t \mathbf{j})$$

Find the smallest values of the elapsed time after which the following quantities are largest and smallest:

(a) the relative distance
(b) the relative speed
(c) the magnitude of the relative acceleration
This problem can be solved with very little mathematics. Draw the velocity and acceleration vectors at the times $t = n\pi/\omega$ where $n = 0, 1, 2, 3, 4$.

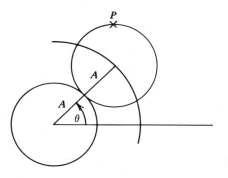

Figure 3.10 Problem 3.12.

3.12 A circle of radius A rolls along another circle of the same radius and fixed in space with constant angular speed ω (see Figure 3.10). Find the position, the velocity, and the speed of a point P on the rim of the revolving circle relative to the center of the fixed circle. At $t = 0$, the coordinates of P relative to the origin of the fixed circle are $(3A, 0)$. (What is the angular speed of rotation of the moving circle in terms of ω?)

3.13 A coordinate system S' oscillates about the origin O of a fixed coordinate system S with constant angular frequency ω and amplitude $\sqrt{2}\,A$ along the straight line $y = x$ (see Figure 3.11). A point P moves on a circle of radius A about the origin O' of S' in counterclockwise direction with angular speed $\omega' = \omega$. Assume that initially $O = O'$ and $y' = 0$. Consider the two cases:

(a) Initially, S' moves into the direction of the first quadrant.
(b) Initially, S' moves into the direction of the third quadrant.
Find the velocity vector $\mathbf{v}$, the speed v, and the acceleration vector $\mathbf{a}$ in each case.

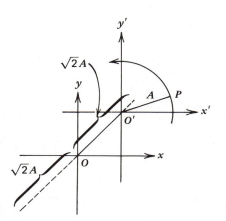

Figure 3.11 Problem 3.13.

chapter

4 Newtonian Dynamics

The greater the obstacle,
the more glory in overcoming it.

MOLIERE

4.1 INTRODUCTION

The fundamental laws (axioms) of classical dynamics were first formulated in a systematic way by Isaac Newton in his *Principia*. They will be introduced and discussed in the next section. We have quickly accepted Newton's laws because they are in a simple form, admit a direct interpretation (except perhaps for the notion of inertia), and are observed in many instances of everyday life. This is in contrast to the basic laws of quantum mechanics which, at first glance, appear strange and not immediately comprehensible because they are remote from everyday experience.

Three different but equivalent formulations of classical dynamics have been developed: Newtonian dynamics, which admits the most direct interpretation of the equations of motion; Lagrangian dynamics, which provides a straightforward procedure for setting up the equations of motion; and Hamiltonian dynamics, which is suited to expose the intrinsic symmetries of the system. If a problem of mechanics cannot be solved (exactly) in any one of these formalisms, it cannot be solved (exactly) in either of the other formalisms. But depending on the questions asked, each formalism may yield at least partial answers. Hamiltonian and Lagrangian dynamics are important not only because of their mathematical simplicity and elegance, but they are also the starting points for statistical mechanics, quantum mechanics and, perhaps, quantum field theory.

4.2 NEWTON'S LAWS

Any dynamical theory must postulate the rules connecting:
(**a**) the characteristic properties of the system (object),
(**b**) the dynamical quantity describing the action on the system, and
(**c**) the reference frame.

In Newtonian mechanics the connection is accomplished by the following laws (axioms):

1 Every particle persists in a state of rest or uniform motion on a straight line (constant velocity) unless it is compelled to change that state by forces impressed on it.

2 The rate of change of the velocity of a particle is proportional to the applied force and points in the direction in which the force acts.

3 Forces always appear in pairs. The mutual actions are always equal in magnitude and opposite in direction.

Although everyone agrees with the mathematical form of Newton's laws (outlined below), different interpretations and logical constructions of the foundations of classical mechanics are possible depending on the choice of the fundamental physical quantities.[1] The following exposition represents one possible line of thought:

We assume that force is a fundamental quantity. This is suggested by the fact that the term *force* appears in all three of Newton's laws without further explanation. Because everyone has experienced force many times and in many forms, conceptional difficulties should not be encountered. However, as shown in the "common sense" examples of Chapter 1 (Questions 1.4, 1.5, and 1.6, which are all centered around the notion of force), the concept of force seems to be much harder to comprehend than one might at first admit. This explains why the dynamical laws of mechanics were discovered only a few hundred years ago, whereas, for example, much of plane geometry had been known for nearly 2000 years before Newton. The invention of calculus by Newton and Leibniz and the introduction of the scientific method by Galilei were also crucial for the development of mechanics.

We shall interpret all three of Newton's laws as dynamical laws because they contain the term *force*. Without specific forces, these laws provide only the empty framework of the theory of mechanics. However, the framework itself is very restrictive, because the laws postulate that the environment influences the position and velocity of a particle via forces only.

The first law refers to the motion of a particle in the absence of forces. Particles or systems that are not acted upon by forces are called free or isolated. Because "uniform motion on a straight line" must be interpreted as constant velocity, the first law may be stated in the preliminary kinematical form

$$\mathbf{v} = \text{constant (for an isolated particle)} \tag{4.1}$$

We shall return to the dynamical formulation of this law after discussing the second law.

The question now arises: What is the reference frame relative to which the first law holds? We call it the inertial frame, and the first law thus implicitly postulates the existence of the inertial frame. More specifically, an inertial frame (in the Newtonian sense) is that reference frame in which Newton's laws and the gravitational force law[2] holds. However, forces that are proportional to the

---

[1]The situation is similar in quantum mechanics. No one questions the validity of the mathematical formalism, but the discussion about the interpretation of quantum mechanics is ongoing.
[2]An inertial frame in the Einsteinian sense is that reference frame in which all forces, including the gravitational force, are negligible.

masses of the objects and can be changed by accelerations and rotations of the reference frame (e.g., the centrifugal force) must not be present or must be negligibly small. A laboratory table may serve as an inertial frame for physical systems in its neighborhood, but it cannot be an inertial frame for our galaxy. This example shows that the size of an inertial frame is necessarily limited.

It also follows from the first law that the existence of one inertial frame implies the existence of an infinite number of other inertial frames. Consider, namely, an object that moves with constant velocity relative to an *inertial frame* S_1. Let S_2 be any nonrotating *reference frame* moving at constant velocity relative to S_1. Then by the transformation equation (3.42) for relative motion, the velocity of the object relative to S_2 is constant, and, by the first law, S_2 must be an inertial frame also.

Newton believed that absolute space could serve as the (preferred) reference frame. His notion of absolute space[3] is that of a container which exists independently of all material objects and in which all objects move. Space is absolute in the sense that it allows us to determine whether an object is moving or absolutely at rest. But because there is no criterion that determines when a reference frame is in uniform motion or at rest, absolute space cannot be detected, and the idea of absolute space as a reference frame has to be abandoned.

We now turn to the second law. Because it contains the phrase "direction of the force," the force must be a vector that we denote by $\mathbf{F}$. The second law then reads

$$m\dot{\mathbf{v}} = \mathbf{F} \tag{4.2}$$

where m is a proportionality constant that is called inertial mass, or just mass. In order to provide an operational definition for mass the third law must be invoked (see below). The second law postulates the relation between the force applied to the particle and the change of motion of the particle expressed by the product of mass and velocity. If we define momentum by

$$\mathbf{p} = m\mathbf{v}$$

Equation 4.2 may be written

$$\dot{\mathbf{p}} = \mathbf{F} \tag{4.3}$$

which is Newton's original version of the second law. Equation 4.3 is more general than Eq. 4.2 because it includes the case of particles with variable mass (e.g., a rocket). An alternate form of (4.2) is

$$m\ddot{\mathbf{r}} = \mathbf{F} \tag{4.4}$$

If $\mathbf{F}$ is a known function of $\mathbf{r}$, $\dot{\mathbf{r}}$ and t, then Eq. 4.4 is a second-order differential equation for the position $\mathbf{r}$ as function of the time t.

We need to clarify the relation between the first and the second law. If $\mathbf{F} = 0$, from (4.3) we obtain $\dot{\mathbf{p}} = 0$, which yields

$$\mathbf{p} = m\mathbf{v} = \text{constant (for an isolated particle)} \tag{4.5}$$

---

[3]For a fascinating account of the many philosophies of space in general and a thorough discussion of Newton's absolute space in particular, see M. Jammer, *Concepts of Space*, Harvard University Press, Cambridge, Mass.: 1954.

For m = constant, this equation is equivalent to (4.1), and we may consider (4.5) as the dynamical form of the first law. But does the second law then imply the first law? The answer is "no" because the first law is necessary to define the inertial frame without which (4.2) could not be stated. The above conclusion merely expresses the fact that the first two laws are compatible with each other when they are to hold in the same inertial frame.

The second law is also invariant under the Galilei transformation (3.43) if the assumption is made that the force remains the same in all inertial frames.

The third law informs us about the origin of the forces and some of their general properties. Forces are manifestations of interactions. Four fundamental interactions are known: The two long-range gravitational and electromagnetic interactions, and the two short-range weak and strong interactions that appear in nuclear and elementary particle physics so that Newton's mechanics are not relevant to them. All macroscopic forces that are not gravitational are electromagnetic forces or of electromagnetic origin, for example, the tension in a rope or the force exerted by a spring. The general forces of macroscopic physics must be known independently of and in addition to the three laws of Newton. They may be obtained from observation or calculated from the laws of the fundamental interactions.

Because forces result from interactions, they must appear in pairs. If two systems A_1 and A_2 interact with each other, then the force $\mathbf{F}_{12}$ of A_1 on A_2 and the force $\mathbf{F}_{21}$ of A_2 on A_1 satisfy the condition

$$\mathbf{F}_{12} = -\mathbf{F}_{21}, \tag{4.6a}$$

which is called the weak form of Newton's third law. It is only satisfied if the transmission of forces is instantaneous. Because the transmission of any signal over a finite distance requires a finite time, as we know now from the theory of relativity, the third law is only an approximation. For this reason, the exact mathematical treatment of several interacting particles in relativity poses difficult problems.

Often the interacting systems are particles located at $\mathbf{r}_1$ and $\mathbf{r}_2$, respectively, and that the force $\mathbf{F}_{12}$ (or $\mathbf{F}_{21}$) is parallel to $\mathbf{r}_1 - \mathbf{r}_2$, that is, points in the direction of the line joining the two particles. Then we have

$$\mathbf{F}_{12} = -\mathbf{F}_{21} \quad \text{parallel to} \quad \mathbf{r}_1 - \mathbf{r}_2 \tag{4.6b}$$

which is called the strong form of Newton's third law. The static forces of electricity and gravitation and the tension in a string are examples of forces that satisfy the strong form of the third law.

We now consider two objects A_0 and A_1 which are assumed to interact with each other. The exact form of the force need not be known. A_0 is an object having the standard (or reference) mass m_0, whereas the mass m_1 of the object A_1 is to be determined. From Eqs. 4.6a and 4.4 we obtain

$$m_1 \ddot{\mathbf{r}}_1 = \mathbf{F}_{01} = -\mathbf{F}_{10} = -m_0 \ddot{\mathbf{r}}_0$$

or

$$m_1 = m_0 \frac{a_0}{a_1} \tag{4.7}$$

where $a_0 = |\ddot{\mathbf{r}}_0|$ and $a_1 = |\ddot{\mathbf{r}}_1|$. The accelerations a_0 and a_1 can be measured, and (4.7) guarantees a meaningful operational definition of the unknown mass m_1. Incidentally, Newton defined mass as the product of density and volume, but he did not give a further explanation of density.

Because by Eq. 4.7 masses are inversely proportional to the accelerations if the force is held constant, mass may be understood as a measure of inertia, the resistance of the object against acceleration when a force is applied to it. The explanation of the origin of inertia goes beyond the content of Newton's laws. Newton believed that absolute space causes the bodies to possess inertia. This explanation is strange in view of the third law: Absolute space acts upon the objects but is itself not acted upon by the bodies. (The asymmetry is removed in general relativity, Einstein's theory of gravitation. Masses cause flat space to curve, and curved space produces the gravitational pull between the masses.) Another explanation of inertia was proposed by E. Mach: The inertia of a body is the result of all interactions of the body with all the other objects in the universe.[4]

We close this section with a brief summary.
Chosen fundamental quantity: Force.

First law: $\mathbf{v} = $ constant (for an isolated particle)
The existence of an inertial frame is postulated.

Second law: $\dot{\mathbf{p}} = \mathbf{F}$
The relation between $\dot{\mathbf{p}}$ and $\mathbf{F}$ is postulated.

Third law: $\mathbf{F}_{12} = -\mathbf{F}_{21}$
Some general properties of forces are postulated.

The second and the third law imply the quantitative definition of inertial mass: $m = m_0 \, (a_0/a)$ where m_0 is a standard mass.
The explicit form of the force must be known (force laws).

4.3   CONSEQUENCES OF THE SECOND LAW

4.3.1   The Impulse-Momentum and the Work-Energy Theorems

It was pointed out in the previous section that the differential equation (4.4) (Newton's second law) can only be solved for the position coordinates of a particle as functions of the time if the exact mathematical form of the force $\mathbf{F}$ is known. We assume that the force $\mathbf{F}$ depends only on the position and velocity

[4]For detailed discussion of Mach's principle, see R. H. Dicke, "The Many Faces of Mach," in *Gravitation and Relativity,* Hong-Yee Chiu and E. F. Hoffmann (eds.), Benjamin, New York: 1964.

coordinates of all particles involved and on the time:[5]

$$\mathbf{F}_i = \mathbf{F}_i(\mathbf{r}_1, \mathbf{r}_2, \cdots \mathbf{r}_N, \dot{\mathbf{r}}_1, \dot{\mathbf{r}}_2, \cdots \dot{\mathbf{r}}_N, t)$$

which in the case of one particle reduces to

$$\mathbf{F} = \mathbf{F}(\mathbf{r}, \dot{\mathbf{r}}, t) = \mathbf{F}(x, y, z, \dot{x}, \dot{y}, \dot{z}, t)$$

Subclasses of forces are[6]

(a) $\mathbf{F} = \mathbf{F}(t)$ for example, the periodic force $F = F_0 \sin \omega t,$

(b) $\mathbf{F} = \mathbf{F}(\dot{\mathbf{r}})$ for example, the frictional force $F = -kv$

(c) $\mathbf{F} = \mathbf{F}(\mathbf{r})$ for example, the gravitational force $F = Gm_1m_2r^{-2}$

Even for forces of the simple form $\mathbf{F}(\dot{\mathbf{r}})$ or $\mathbf{F}(\mathbf{r})$, general procedures for solving Newton's second law do not exist. In Section 4.5 we shall present various methods for solving the differential equation when special types of forces are considered.

We now assume that the coordinates of a mass point moving under the influence of a force $\mathbf{F}(\mathbf{r}, \dot{\mathbf{r}}, t)$ are known for all values of the time t. The coordinates of $\mathbf{r} = \mathbf{r}(t)$ may be obtained from the differential equation (4.4) (at least in principle) or by performing measurements. It is then possible to consider the force as dependent on the time alone,

$$\mathbf{F}[\mathbf{r}(t), \dot{\mathbf{r}}(t), t] \to \mathbf{F}(t) \tag{4.8}$$

or as dependent on the position coordinates alone,

$$\mathbf{F}[\mathbf{r}(t), \dot{\mathbf{r}}(t), t] \to \mathbf{F}(\mathbf{r}) \tag{4.9}$$

In the latter case choose, for example, $x = x(t)$, which may be inverted to $t = t(x)$, and both t and $\mathbf{r}(t(x))$ are position dependent. On the other hand, if $\mathbf{r}(t)$ is known, $\mathbf{F}$ is no longer needed except to *verify* equations of motion or conservation laws.

We define two important quantities. The impulse $\mathcal{I}$ is the time integral of $\mathbf{F}$:

$$\mathcal{I} = \int_{t_1}^{t_2} \mathbf{F}\, dt \tag{4.10}$$

Its existence is guaranteed by the relation (4.8). The work W done by the force on the particle is the line integral

$$W = \int_{P_1}^{P_2} \mathbf{F} \cdot d\mathbf{r} \tag{4.11}$$

[5]In electromagnetic theory some forces depend on $\dot{\mathbf{r}}$ and $\ddot{\mathbf{r}}$.
[6]See Appendix 2.

where C denotes the *actual* path of the particle. The existence of W is guaranteed by the relation (4.9). With the time t as the path parameter, the integral (4.11) may be brought into the form

$$W = \int_{t_1}^{t_2} \mathbf{F}(t) \cdot \frac{d\mathbf{r}(t)}{dt} dt = \int_{t_1}^{t_2} P(t) \, dt \tag{4.12}$$

where

$$P(t) = \mathbf{F} \cdot \mathbf{v} \tag{4.13}$$

is the instantaneous power delivered to the mass point. It then follows that

$$P = \frac{dW}{dt} \tag{4.14}$$

Another quantity that will be needed below is the kinetic energy T defined by

$$T = \tfrac{1}{2} m\dot{\mathbf{r}} \cdot \dot{\mathbf{r}} = \tfrac{1}{2} m v^2 \tag{4.15}$$

We are now prepared to derive two important conclusions from Newton's second law. If we integrate the equation

$$\frac{d\mathbf{p}}{dt} = \mathbf{F}$$

with respect to t we obtain

$$\Delta\mathbf{p} = \mathbf{p}(t_2) - \mathbf{p}(t_1) = \int_{t_1}^{t_2} \frac{d\mathbf{p}}{dt} dt = \int_{t_1}^{t_2} \mathbf{F} \, dt$$

or

$$\Delta\mathbf{p} = \mathcal{J} \tag{4.16}$$

which is called the impulse-momentum theorem.

If we evaluate the inner product of $\dot{\mathbf{p}}$ and $d\mathbf{r}$, we obtain, using the identity (3.6),

$$\dot{\mathbf{p}} \cdot d\mathbf{r} = m\frac{d\mathbf{v}}{dt} \cdot d\mathbf{r} = m\frac{d\mathbf{v}}{dt} \cdot \frac{d\mathbf{r}}{dt} dt$$

$$= m\frac{d\mathbf{v}}{dt} \cdot \mathbf{v} \, dt = \frac{d}{dt}\left(\frac{1}{2} m v^2\right) dt = dT$$

Therefore, the line integral of Eq. 4.3

$$\int_{P_1 \atop C}^{P_2} \dot{\mathbf{p}} \cdot d\mathbf{r} = \int_{P_1 \atop C}^{P_2} \mathbf{F} \cdot d\mathbf{r}$$

is equal to

$$\Delta T = T_2 - T_1 = W \qquad (4.17)$$

which is called the work-energy theorem.

Quantities such as **p** and T, which describe properties of the system (mass point) itself, are called state variables. Quantities such as $\mathcal{I}$ and W, which describe certain actions (caused by the force) on the system are, therefore, not state variables.

4.3.2 The Rotational Impulse-Momentum and Work-Energy Theorems

Important conclusions may be drawn from Newton's second law for circular motion. We define the two axial vectors torque **N** and angular momentum **l** by

$$\mathbf{N} = \mathbf{r} \times \mathbf{F} \qquad (4.18)$$

and

$$\mathbf{l} = \mathbf{r} \times \mathbf{p} = m\mathbf{r} \times \mathbf{v} \qquad (4.19)$$

Both **N** and **l** are coordinate system-dependent vectors because they contain the position vector **r** explicitly. Taking the time derivative of **l** we obtain

$$\dot{\mathbf{l}} = \dot{\mathbf{r}} \times \mathbf{p} + \mathbf{r} \times \dot{\mathbf{p}}$$
$$= \dot{\mathbf{r}} \times m\dot{\mathbf{r}} + \mathbf{r} \times \mathbf{F}$$

The first term on the right-hand side vanishes identically, and the second term is equal to the torque. The resulting equation

$$\dot{\mathbf{l}} = \mathbf{N} \qquad (4.20)$$

is the rotational equivalent to the equation of motion (4.3).

If **v** is replaced by $\boldsymbol{\omega} \times \mathbf{r}$, see (3.37), both the angular momentum and the kinetic energy may be written in terms of **r** and $\boldsymbol{\omega}$:

$$\mathbf{l} = m\mathbf{r} \times \mathbf{v} = m\mathbf{r} \times (\boldsymbol{\omega} \times \mathbf{r}) = mr^2\boldsymbol{\omega} - m(\mathbf{r} \cdot \boldsymbol{\omega})\mathbf{r}, \qquad (4.21)$$

$$T = \tfrac{1}{2}mv^2 = \tfrac{1}{2}m(\boldsymbol{\omega} \times \mathbf{r}) \cdot (\boldsymbol{\omega} \times \mathbf{r}) = \tfrac{1}{2}m\left[r^2\omega^2 - (\mathbf{r} \cdot \boldsymbol{\omega})^2\right] \qquad (4.22)$$

The structure of these results is exhibited with the matrix (tensor) of the moment of inertia

$$I_{kl} = m\left(r^2\delta_{kl} - x_k x_l\right) \qquad (4.23)$$

(All we need to know here is that the moment of inertia is represented by a symmetric 3×3 matrix. We are not interested in the transformation properties of the tensor I.) The coordinates of **l** are then given by

$$l_k = \sum_{l=1}^{3} I_{kl}\omega_l,$$

and the kinetic energy becomes

$$T = \frac{1}{2} \sum_{k=1}^{3} \sum_{l=1}^{3} \omega_k I_{kl} \omega_l$$

These equations read in matrix form

$$\mathbf{l} = \mathsf{I}\omega \tag{4.24}$$

and

$$T = \frac{1}{2}\omega^T \mathsf{I}\omega \tag{4.25}$$

Observe that $\mathbf{l}$ and ω are not parallel vectors.

The expressions (4.21) and (4.22) for $\mathbf{l}$ and T take on an even simpler form if $\mathbf{r}$ is perpendicular to ω. This can always be achieved by a transformation of the coordinate system. Because $\mathbf{r} \cdot \omega = 0$, we have

$$\mathbf{l} = mr^2\omega = I\omega \tag{4.26}$$

and

$$T = \frac{1}{2}mr^2\omega^2 = \frac{1}{2}I\omega^2 \tag{4.26a}$$

where

$$I = mr^2 \tag{4.27}$$

is the moment of inertia of the mass point about the axis of rotation. The vectors ω and $\mathbf{l}$ are now parallel because I is a scalar. Using (4.27) we may write the equation of motion (4.20) in the form

$$(I\omega)^{\cdot} = \mathbf{N},$$

or, if I is time independent, that is, if $\dot{r} = 0$,

$$I\dot{\omega} = I\alpha = \mathbf{N} \tag{4.28}$$

This equation is formally equivalent to (4.2).

We now proceed to derive the rotational equivalent of the impulse-momentum theorem and the work-energy theorem.

If we integrate Eq. 4.20 with respect to t, we obtain

$$\Delta\mathbf{l} = \mathbf{l}(t_2) - \mathbf{l}(t_1) = \int_{t_1}^{t_2} \mathbf{N}\, dt = \mathcal{G}_{\mathrm{rot}} \tag{4.29}$$

which is the angular impulse-angular momentum theorem. The angular momentum remains constant if $\mathcal{G}_{\mathrm{rot}}$ vanishes. This is obviously true if $\mathbf{N} = 0$, that is, if either $\mathbf{r} = 0$ or $\mathbf{F} = 0$, or if $\mathbf{F}$ and $\mathbf{r}$ are parallel. A force of the form

$$\mathbf{F}(\mathbf{r}) = f(\mathbf{r})\mathbf{e}_r \qquad (4.30)$$

where $\mathbf{e}_r = \mathbf{r}/r$ is the unit vector in radial direction, is called a central force. Because for central forces $\mathbf{r} \times \mathbf{F} = 0$, it follows from the theorem (4.29) that the angular momentum of a mass point remains unchanged if it is acted upon by a central force. Note that $f(\mathbf{r}) = f(x, y, z) \neq f(r)$.

In order to derive the rotational work-energy theorem, we take the inner product of both sides of (4.20) with $\boldsymbol{\omega}$ such that

$$\boldsymbol{\omega} \cdot \dot{\mathbf{l}} = \boldsymbol{\omega} \cdot \mathbf{N} \qquad (4.31)$$

The left side of (4.31) is equal to the time derivative of the rotational kinetic energy,

$$\boldsymbol{\omega} \cdot \dot{\mathbf{l}} = \frac{dT}{dt}$$

The proof of this equation would turn out to be too lengthy if we used the relations (4.21) and (4.23). But we can apply a powerful and very elegant trick: Because the meaning of any physical quantity remains unchanged under a coordinate transformation, we can choose that coordinate system relative to which the quantity assumes the simplest form. It is therefore sufficient for our purpose to use Eqs. 4.26 and 4.26a. Assuming that $\dot{I} = 0$, we obtain

$$\frac{dT}{dt} = \frac{d}{dt}\left(\frac{1}{2}I\omega^2\right) = \omega I\dot{\omega} = \boldsymbol{\omega} \cdot (I\dot{\boldsymbol{\omega}}) = \boldsymbol{\omega} \cdot \dot{\mathbf{l}} \qquad (4.32)$$

The right side of (4.31) is the power delivered to the mass point by the torque $\mathbf{N}$:

$$\boldsymbol{\omega} \cdot \mathbf{N} = \boldsymbol{\omega} \cdot (\mathbf{r} \times \mathbf{F}) = \mathbf{F} \cdot (\boldsymbol{\omega} \times \mathbf{r}) = \mathbf{F} \cdot \mathbf{v} = \frac{dW}{dt} \qquad (4.33)$$

Here we have used (3.37), (4.13), and (4.14). Combining (4.31), (4.32), and (4.33), and integrating with respect to t, we finally get

$$\Delta T = T(t_2) - T(t_1) = \int_{t_1}^{t_2} \boldsymbol{\omega} \cdot \dot{\mathbf{l}} \, dt$$

$$= \int_{t_1}^{t_2} \boldsymbol{\omega} \cdot \mathbf{N} \, dt = \int_{t_1}^{t_2} \frac{dW}{dt} \, dt = W \qquad (4.34)$$

which is the rotational work-energy theorem.

4.4 CONSERVATIVE FORCES AND POTENTIAL ENERGY

We consider a mass point that is moved by a position dependent force $\mathbf{F}(\mathbf{r})$ from the point P_0 to the point P along a path C. The work done by the force *on the* mass point is the line integral

$$W = \int_{P_0 \atop C}^{P} \mathbf{F} \cdot d\mathbf{r} \tag{4.35}$$

which is in general path dependent (see Section 2.6.3). Of great importance in physics are the conservative forces, for which the line integral (4.35) is path independent. Systems that are acted upon by conservative forces only are called conservative systems. Although conservative forces are more *manageable*, others are more widespread in nature. For any two paths C_1 and C_2 (see Figure 4.1) we have

$$\int_{P_0 \atop C_1}^{P} \mathbf{F} \cdot d\mathbf{r} = \int_{P_0 \atop C_2}^{P} \mathbf{F} \cdot d\mathbf{r}$$

If we invert the direction of the path C_2, we obtain

$$\int_{P_0 \atop C_1}^{P} \mathbf{F} \cdot d\mathbf{r} = -\int_{P \atop -C_2}^{P_0} \mathbf{F} \cdot d\mathbf{r}$$

or

$$\oint \mathbf{F} \cdot d\mathbf{r} = 0$$

The line integral of a conservative force along *any* closed path is zero. It then follows from the general theory of line integrals that the differential form $\mathbf{F} \cdot d\mathbf{r}$ is a perfect differential. We denote this differential by $-dV(\mathbf{r})$ and call $V(\mathbf{r})$ the potential energy function. Because $V(\mathbf{r})$ is defined by the differential

$$dV(\mathbf{r}) = -\mathbf{F} \cdot d\mathbf{r} \tag{4.36}$$

the quantity $V(\mathbf{r})$ can be determined relative to a reference value $V(r_{\text{ref}})$ only:

$$V(\mathbf{r}) - V(\mathbf{r}_{\text{ref}}) = -\int_{P_{\text{ref}}}^{P} \mathbf{F} \cdot d\mathbf{r}$$

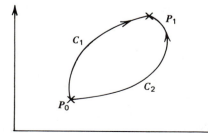

Figure 4.1 Path independence of the work done by conservative forces.

The point of reference P_{ref} is usually chosen in such a way that the function $V(\mathbf{r})$ takes on the simplest mathematical form. In most cases one sets $V(\mathbf{r}_{ref}) = 0$.

The work W can now be expressed by the potential energy:

$$W = \int_{P_0}^{P} \mathbf{F} \cdot d\mathbf{r} = \int_{P_{ref}}^{P} \mathbf{F} \cdot d\mathbf{r} - \int_{P_{ref}}^{P_0} \mathbf{F} \cdot d\mathbf{r}$$

$$= -\left[V(\mathbf{r}) - V(\mathbf{r}_{ref}) \right] + \left[V(\mathbf{r}_0) - V(\mathbf{r}_{ref}) \right]$$

$$= -\left[V(\mathbf{r}) - V(\mathbf{r}_0) \right] = -\Delta V \tag{4.37}$$

The reference value $V(r_{ref})$ has dropped out. If positive work is done by the conservative force, then the potential energy of the system decreases.

Equation 4.36 relates the components of the force to the partial derivatives of the potential energy:

$$\mathbf{F} = -\nabla V \tag{4.38}$$

or

$$F_x = -\frac{\partial V}{\partial x}, \quad F_y = -\frac{\partial V}{\partial y}, \quad F_z = -\frac{\partial V}{\partial z} \tag{4.38a}$$

The force points in the direction opposite to the steepest slope of the potential energy. For the one-dimensional case, $F(x) = -dV(x)/dx$, the relation between F and V is shown in Figure 4.2. If $\mathbf{F} = 0$, the mass point is said to be in equilibrium. The equilibrium is stable if V is a minimum, it is unstable if V is a maximum. In Figure 4.2, the point of unstable equilibrium is A, and the point of stable equilibrium is B.

Equation 4.38 implies that

$$\nabla \times \mathbf{F} = 0 \tag{4.39}$$

or

$$\frac{\partial F_x}{\partial y} = \frac{\partial F_y}{\partial x}, \quad \frac{\partial F_y}{\partial z} = \frac{\partial F_z}{\partial y}, \quad \frac{\partial F_z}{\partial x} = \frac{\partial F_x}{\partial z} \tag{4.40}$$

which is the necessary and sufficient condition for $\mathbf{F}$ to be conservative. A one-dimensional position dependent force $F(x)$ is always conservative.

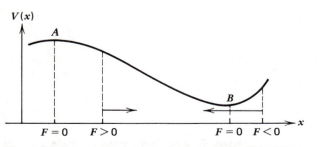

Figure 4.2 The relation between force and potential energy.

Example 4.1

Is $\mathbf{F} = (1/2)y^2\mathbf{i} + xy\mathbf{j} + xz\mathbf{k}$ conservative? By conditions (4.40) we should have

$$\frac{\partial \frac{1}{2}y^2}{\partial y} = \frac{\partial xy}{\partial x}, \quad \frac{\partial xy}{\partial z} = \frac{\partial xz}{\partial y}, \quad \frac{\partial xz}{\partial x} = \frac{\partial \frac{1}{2}y^2}{\partial z}$$

Although the first two equations are identically satisfied, $y = y$ and $0 = 0$, the last equation gives $z = 0$. Hence, the force is not conservative. ∎

Two subclasses of conservative forces are of special importance.

(a) A separable force is a force of the form

$$\mathbf{F}(\mathbf{r}) = F_x(x)\mathbf{i} + F_y(y)\mathbf{j} + F_z(z)\mathbf{k} \tag{4.41}$$

and the potential can be written

$$V(\mathbf{r}) = V_1(x) + V_2(y) + V_3(z) \tag{4.42}$$

(b) In the preceding section we introduced the central force

$$\mathbf{F}(\mathbf{r}) = f(\mathbf{r})\mathbf{e}_r \tag{4.30}$$

which leaves the angular momentum unchanged but which is in general not conservative. However, the central isotropic force that satisfies the stronger condition

$$\mathbf{F}(\mathbf{r}) = f(r)\mathbf{e}_r \tag{4.43}$$

is conservative. The potential energy function

$$V(r) = -\int_{r_0}^r \mathbf{F}(\mathbf{r}') \cdot d\mathbf{r}' = -\int_{r_0}^r f(r')\mathbf{e}_r \cdot d\mathbf{r}' = -\int_{r_0}^r f(r')\, dr'$$

is a function of r only, and it follows that

$$f(r) = -\frac{dV(r)}{dr} \tag{4.44}$$

is an ordinary derivative.

4.5 THE THEOREM OF ENERGY CONSERVATION

The application of the work-energy theorem to conservative systems leads to an important result. If we combine Eqs. 4.17 and 4.37, we obtain

$$\Delta T = W = -\Delta V$$

which can be written as

$$\Delta(T + V) = 0$$

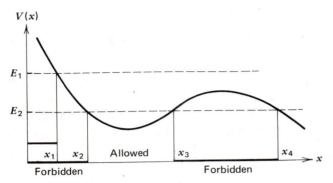

Figure 4.3 Physical and unphysical regions.

or

$$T + V = \text{constant} \qquad (4.45)$$

This equation is known as the theorem of energy conservation. We define the total mechanical energy E by

$$E = T + V \qquad (4.46)$$

Then, (4.45) becomes simply

$$E = \text{constant} \qquad (4.47)$$

Because the work-energy theorem involves one integration over the time t, the constant energy E is often called a first integral of the equation of motion. One should clearly distinguish between the two equations (4.45) and (4.46): whereas (4.45) is a theorem, (4.46) is a definition.

The theorem of energy conservation (4.45) is a direct consequence of Newton's second law applied to conservative systems. Some velocity dependent forces may also lead to a constant value of the total energy. If F is of the form $\mathbf{v} \times \mathbf{b}$, where $\mathbf{b}$ is a constant vector (e.g., the force on an electric charge moving through a constant magnetic field), then $\mathbf{F} \cdot \mathbf{v} = (\mathbf{v} \times \mathbf{b}) \cdot \mathbf{v} = 0$ and $W = 0$. The potential energy does not exist[7] in this case, and the work-energy theorem becomes $\Delta T = 0$ or

$$T = E = \text{constant} \qquad (4.48)$$

The theorem of energy conservation provides us with useful information about the motion of the particle. As V increases, T and the speed v decrease, and vice versa. Because T is never negative the maximum value of V is reached when $T = 0$ or $V = E$. Regions for which $V > E$ are forbidden (or unphysical).

Let us consider a potential energy function shown in Figure 4.3. If $E = E_1$, the mass point may move to any point $x \geqslant x_1$. The unphysical region is $x < x_1$. If

---

[7]See, however, Section 8.10.

$E = E_2$, the mass point will remain either in the interval $x_2 \leqslant x \leqslant x_3$ performing oscillations, or in the region $x \geqslant x_4$. The mass point cannot penetrate the potential barrier between x_3 and x_4. In quantum mechanics, however, there is a positive probability for the particle to move through the potential barrier from one physical region to another while the total energy remains constant.

We conclude this section with a brief summary. If **F** denotes a force, then the following statements are equivalent:

1. **F(r)** is conservative.
2. $\int_{P_1}^{P_2} \mathbf{F} \cdot d\mathbf{r}$ is independent of the path.
3. $\oint \mathbf{F} \cdot d\mathbf{r} = 0$ for any closed path.
4. $\mathbf{F} \cdot d\mathbf{r}$ is a perfect differential.
5. $\nabla \times \mathbf{F} = 0$ everywhere.
6. A (potential) function $V(\mathbf{r})$ exists such that $\mathbf{F} = -\nabla V$.
7. $T + V = \text{constant}$ (this statement also involves the equations of motion).

4.6 METHODS FOR SOLVING NEWTON'S EQUATIONS

4.6.1 Initial Conditions

In the case of one mass point, Newton's second law (4.4) represents a set of three second-order differential equations. Each second-order differential equation has two linearly independent solutions and admits two arbitrary integration constants. The general solution $\mathbf{r}(t)$ describing the motion of the particle in three dimensions contains six arbitrary constants. Conditions imposed on these constants single out one solution that describes the actual orbit of the particle. Natural choices are the six initial values of the position and velocity vectors:

$$at \quad t = t_0: \quad \mathbf{r}(t_0) = \mathbf{r}_0, \quad \mathbf{v}(t_0) = \mathbf{v}_0$$

If the initial values $\mathbf{r}_0$ and $\mathbf{v}_0$ are prescribed, then the position vector $\mathbf{r}$ is exactly known at all times. Classical mechanics, therefore, is a deterministic theory in which probabilities do not occur.

There are other possibilities to fix the arbitrary integration constants. One may, for example, require that certain conserved dynamical quantities of the system such as the total energy and/or the angular momentum assume certain values. We must keep in mind that the number of independent constants has to be six in the three-dimensional space. Specifying only total energy and angular momentum (1 + 3 constants) would be insufficient to describe an orbit completely.

We now turn to the main problem, namely, to solve the equation of motion (4.4) for special types of forces. To simplify the equations we set $t_0 = 0$ without loss of generality.

4.6.2 Time Dependent Forces

For $\mathbf{F} = \mathbf{F}(t)$ the integration of Eq. 4.4 is straightforward. The impulse-momentum theorem (4.15) can be applied, which gives us the first integral of (4.4):

$$m \int_0^t \frac{d\mathbf{v}}{dt}\, dt = m[\mathbf{v}(t) - \mathbf{v}(0)] = \int_0^t \mathbf{F}(t')\, dt'$$

The second integration yields

$$\int_0^t \mathbf{v}(t'')\, dt'' = \int_0^t \frac{d\mathbf{r}(t'')}{dt''}\, dt''$$

$$= \mathbf{r}(t) - \mathbf{r}(0) = \frac{1}{m}\int_0^t \left[\int_0^{t''} \mathbf{F}(t')\, dt' + m\mathbf{v}(0)\right] dt''$$

such that

$$\mathbf{r}(t) = \frac{1}{m}\int_0^t \left[\int_0^{t''} \mathbf{F}(t')\, dt'\right] dt'' + \mathbf{v}_0 t + \mathbf{r}_0 \tag{4.49}$$

The integration constants are $\mathbf{v}_0 = \mathbf{v}(0)$ and $\mathbf{r}_0 = \mathbf{r}(0)$. For the special case of a constant force $\mathbf{F} = \mathbf{F}_0$, we obtain the familiar result

$$\mathbf{r}(t) = \frac{1}{2}\frac{\mathbf{F}_0}{m}t^2 + \mathbf{v}_0 t + \mathbf{r}_0 \tag{4.50}$$

Example 4.2

Find $x(t)$ if $F(t) = bt^2$.
We evaluate $\int_0^t bt'^2\, dt' = \frac{1}{3}bt^3$ and $\int_0^t \frac{1}{3}bt''^3\, dt'' = \frac{1}{12}bt^4$, and the answer is $x(t) = \frac{1}{12}\frac{b}{m}t^4 + v_0 t + x_0$. ■

4.6.3 Velocity Dependent Forces

We first consider the one-dimensional case $F = F(v)$. This problem is easily solved if we follow the indicated steps:

Step 1. We rewrite $m\dot{v} = F(v)$ in the form

$$\frac{dv}{F(v)} = \frac{1}{m}dt$$

and integrate

$$\int_{v_0}^v \frac{dv'}{F(v')} = g(v) - g(v_0) = \frac{1}{m}\int_0^t dt' = \frac{1}{m}t \tag{4.51}$$

where $g(v)$ is the integral of the function $1/F(v)$. Then t is known as a function of v: $t = t(v)$.

Step 2. Assuming that $t(v)$ can be inverted, we calculate $v = v(t)$.

Step 3. If $v(t)$ is known, integration over t yields

$$x(t) = \int_0^t v(t')\, dt' + x_0 \tag{4.52}$$

Example 4.3

Let $F(v) = -bv$. An example of this force acting opposite the motion is the frictional force of air on a moving object at small velocities.

Step 1.

$$\int_{v_0}^v \frac{dv'}{F(v')} = -\frac{1}{b} \int_{v_0}^v \frac{dv'}{v'} = -\frac{1}{b} \ln \frac{v}{v_0} = \frac{1}{m} t$$

Step 2. Solving for v one obtains

$$v(t) = v_0 e^{-(b/m)t}$$

Step 3.

$$x(t) = \int_0^t v_0 e^{-(b/m)t'}\, dt' = \frac{mv_0}{b} \left[1 - e^{-(b/m)t}\right] + x_0$$

As t approaches infinity, the particle comes to rest at $x - x_0 = mv_0/b$. The distance traveled is proportional to the initial momentum and inversely proportional to the constant b, which describes the strength of the friction. ∎

We now turn to velocity dependent forces in two and three dimensions. Only three classes of forces with relatively simple structure will be discussed.

If the force is separable, that is, of the form

$$\mathbf{F}(\mathbf{v}) = F_x(v_x)\mathbf{i} + F_y(v_y)\mathbf{j} + F_z(v_z)\mathbf{k}$$

then the equation of motion

$$m \frac{d\mathbf{v}}{dt} = \mathbf{F}(\mathbf{v})$$

reduces to three independent differential equations of the type discussed above.

If the force has the property that $\mathbf{F} \cdot \mathbf{v}$ is a scalar function of the speed alone, $\mathbf{F} \cdot \mathbf{v} = f(v) = h(v^2)$, then

$$\dot{\mathbf{p}} \cdot \mathbf{v} = \frac{d}{dt}\left(\frac{1}{2}mv^2\right) = \mathbf{F} \cdot \mathbf{v} = h(v^2)$$

from which follows

$$\int_{u_0}^{u} \frac{du'}{h(u')} = \frac{2}{m} \int_0^t dt' = \frac{2}{m} t,$$

where $u = v^2$. The function $t = t(u) = t(v^2)$ may be inverted to $v = v(t)$. Performing another integration we obtain

$$s(t) = \int_0^t v(t') \, dt' + s_0,$$

which is the path coordinate (arc length) of the mass point. Only a partial answer has been obtained in this case.

If the force obeys $\mathbf{F} \cdot \mathbf{v} = 0$ ($\mathbf{F}$ is perpendicular to $\mathbf{v}$), the speed of the particle remains unchanged due to (4.48). Important examples are the magnetic force $\mathbf{F}_m$ on a particle with electric charge q moving in a magnetic field $\mathbf{B}$,

$$\mathbf{F}_m = q\mathbf{v} \times \mathbf{B} \tag{4.53}$$

and the fictitious Coriolis force

$$\mathbf{F}_{\text{Cor}} = -2m(\boldsymbol{\omega} \times \mathbf{v}), \tag{4.54}$$

which appears in rotating reference frames. For these two forces that depend on the coordinates of $\mathbf{v}$ linearly, the method of elimination of variables by substitution may be useful to obtain differential equations that depend on one variable only. An example will be discussed in Section 5.8.

4.6.4 Position Dependent Forces

We begin again with the one-dimensional case. For forces $F = F(x)$ the potential energy function $V(x)$ always exists, and the theorem of energy conservation can be written in the form

$$T = \tfrac{1}{2}mv^2 = E - V(x) \tag{4.55}$$

The reference value $V(x_{\text{ref}})$ of the potential energy also enters the total energy E; hence, it cancels in the kinetic energy (4.55) which is made unique by its vanishing for $v = 0$. Depending on the problem considered, the value of the energy constant E may be prescribed or obtained from the initial conditions imposed on the position coordinate x and the velocity v. If we solve (4.55) for v, we obtain

$$v = \frac{dx}{dt} = \pm \sqrt{\frac{2}{m}[E - V(x)]} \tag{4.56}$$

Separation of the variables and integration leads to

$$t = \int_0^t dt' = \pm \int_{x_0}^{x} \frac{dx'}{\sqrt{\dfrac{2}{m}[E - V(x')]}} \tag{4.57}$$

The time t is then found as a function of the position x. The correct sign in (4.57) is obtained from the sign of the velocity (4.56). If the particle moves to the right (or left), its velocity is positive (or negative). Assuming that the equation $t = t(x)$ can be inverted, we finally calculate $x = x(t)$.

Let us summarize. In order to solve the equation $m\ddot{x} = F(x)$, we perform the following operations:

Step 1. Calculate $V(x)$ from $F(x)$.

Step 2. Apply the initial conditions to calculate E.

Step 3. Evaluate the integral (4.57).

Step 4. Transform $t = t(x)$ into $x = x(t)$.

We see that the energy equation is a powerful tool for solving the equation of motion. This is generally true for all conservation laws which provide not only qualitative but also quantitative information about the system.

Example 4.4

Solve the equation of motion for the force $F = -k/x^3$. The particle is subject to the initial conditions $x(0) = x_0$ and $v(0) = 0$.

Step 1. $\displaystyle V(x) = -\int^x F(x')\,dx' = \int^x \frac{k}{x'^3}\,dx' = -\frac{k}{2}\frac{1}{x^2}$

Step 2. $\displaystyle E = T(v_0) + V(x_0) = \frac{1}{2}mv_0^2 - \frac{k}{2}\frac{1}{x_0^2} = -\frac{k}{2}\frac{1}{x_0^2}$

Step 3. $\displaystyle E - V(x) = \frac{k}{2}\left(\frac{1}{x^2} - \frac{1}{x_0^2}\right)$

The velocity is negative, and we must choose the "$-$" sign in (4.57). Thus,

$$t = -\int_{x_0}^x \frac{dx'}{\sqrt{\dfrac{2}{m}\dfrac{k}{2}\left(\dfrac{1}{x'^2} - \dfrac{1}{x_0^2}\right)}} = -\sqrt{\frac{m}{k}}\,x_0\int_{x_0}^x \frac{x'\,dx'}{\sqrt{x_0^2 - x'^2}} = \sqrt{\frac{m}{k}}\,x_0\sqrt{x_0^2 - x^2}$$

Step 4. $\displaystyle x(t) = x_0\sqrt{1 - \frac{k}{mx_0^4}t^2}$

The mass point reaches the origin $x = 0$ at $t = \sqrt{m/k}\,x_0^2$. ∎

In three dimensions, a few classes of position dependent forces $\mathbf{F}(\mathbf{r})$ exist that admit straightforward procedures for solving the equations of motion. These forces possess a large degree of symmetry, and it is a very fortunate coincidence that nature prefers these forces over other forces with a more complicated structure.

If the force is separable (see Eq. 4.41), the equations of motion reduce to three independent one-dimensional equations which can be solved separately following the procedures outlined at the beginning of this section.

Solution methods for problems involving central isotropic forces (Eq. 4.43) will be discussed in Chapter 6.

A situation often arises where several forces $\mathbf{F}_1, \mathbf{F}_2, \ldots$, acting simultaneously on a particle, obey $F_1 \gg F_2 \gg \cdots$. Here, the methods of perturbation theory may be applied. One first solves the equation of motion with the force $\mathbf{F}_1$ alone and modifies the solution when the force $\mathbf{F}_2$ is included. It is assumed that the small force $\mathbf{F}_2$ causes only a small perturbation of the orbit.

Sometimes it is desired to know the orbit $y = y(x)$ of a particle rather than the instantaneous position $\mathbf{r}(t)$. It is possible under certain conditions to eliminate the time variable and to derive a differential equation involving the space coordinates only.

The methods described above for solving the equations of motion were based mainly on physical principles and arguments. It should be clear that purely mathematical methods developed in the theory of differential equations can also be used, and often this is the only way to proceed. The physical aspects of the problem should always be kept in mind, regardless of which method is applied.

QUESTIONS

4.1 Discuss the meaning of axioms. Compare the views expressed in this book with the following statement:[8]

As Mach showed, the Newtonian foundations of mechanics can be remodeled in such a way that one does not begin with definitions, but with assertions (axioms), which also suffice to define *the fundamental concepts of force and mass. Then, all one need add to the axioms is explanation of words serving as verbal abbreviations.*

The following quotation[9] is also of interest in this connection:

It should be emphasized that the mathematical formulation of the first law [of thermodynamics] contains three related ideas: (1) the existence of an internal-energy function; (2) the principle of the conservation of energy; (3) the definition of heat as energy in transit by virtue of a temperature difference.

Note that this statement contains the words *existence*, *principle*, and *definition*.

4.2 Draw the directions of the angular momentum and torque vectors for the mass point in circular motion relative to points A and B, respectively, shown

[8]Richard von Mises, "Mathematical Postulates and Human Understanding" in *The World of Mathematics*, Vol. 3, J. R. Newman (ed.), Simon & Schuster, New York: 1956, p. 1727.
[9]M. W. Zemansky, *Heat and Thermodynamics*, 5th ed., McGraw-Hill, New York: 1968, p. 79.

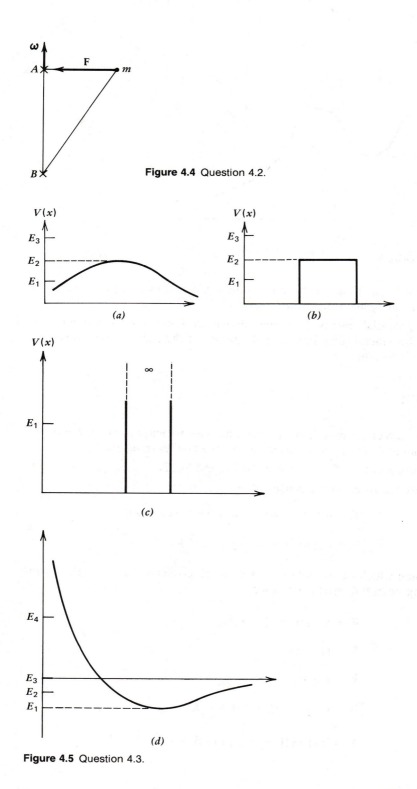

Figure 4.4 Question 4.2.

(a)

(b)

(c)

(d)

Figure 4.5 Question 4.3.

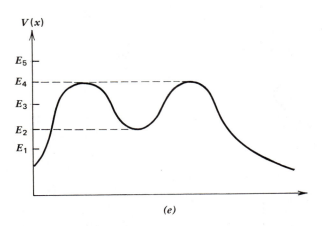

Figure 4.5 (*Continued*)

in Figure 4.4. The force on the mass point is directed toward point A in both cases. What are your conclusions?

4.3 For the potential energy functions shown in Figures 4.5a through 4.5e describe the general (one-dimensional) motion of the particle. Investigate the cases E_i = constant.

PROBLEMS

4.1 Show by direct computation that the impulse-momentum theorem and the work-energy theorem are invariant under the Galilei transformation.

4.2 Derive the equation $\dot{T} = \boldsymbol{\omega} \cdot \mathbf{l}$ from (4.21) and (4.22).

4.3 When are the forces listed conservative?

(a) $\qquad\qquad \mathbf{F} = \mathbf{a} \times (\mathbf{r} \times \mathbf{b}) \qquad$ **a** and **b** are constant

(b) $\qquad\qquad \mathbf{F} = axyze^{x^2}\mathbf{i} + bze^{x^2}\mathbf{j} + cye^{x^2}\mathbf{k}$

4.4 Determine which of the following forces are conservative. Find the corresponding potential energy functions.

(a) $\qquad\qquad \mathbf{F} = ax^2\mathbf{i} + by^2\mathbf{j} + cz^2\mathbf{k}$

(b) $\qquad\qquad \mathbf{F} = y\mathbf{i} - x\mathbf{j}$

(c) $\qquad\qquad \mathbf{F} = \mathbf{r} \times \mathbf{b}$

(d) $\qquad\qquad \mathbf{F} = k\dfrac{z}{y^2}\mathbf{i} - 2k\dfrac{xz}{y^3}\mathbf{j} + k\dfrac{x}{y^2}\mathbf{k}$

(e) $\qquad\qquad \mathbf{F} = \sin(yz)\mathbf{i} + xz\cos(yz)\mathbf{j} + xy\cos(yz)\mathbf{k}$

4.5 Show that a conservative central force is central isotropic.

4.6 Show by direct computation that the central force $\mathbf{F} = xyze_r$ is not conservative.

4.7 What is the most general form of a separable and central force?

4.8 A particle is subjected to the force $F = -ae^{bt}$. Find v and x as functions of t. The initial values are $x(0) = 0$ and $v(0) = v_0$. When and where will the particle come to rest? Simplify the answer for small v_0.

4.9 A particle is subjected to the force $F = F_0 e^{-\lambda t}\sin(\omega t + \delta)$. Find v and x as functions of time. What is the terminal velocity? (*Hint*: $\sin\alpha = (e^{i\alpha} - e^{-i\alpha})/2i$)

4.10 An object falls toward the earth under the influence of the constant gravitational force mg and a force of friction proportional to the speed v. Find the terminal speed of the particle.

4.11 A particle is thrown vertically into the air where it experiences a frictional force proportional to v^2 (constant b).

(a) Set up the equation of motion.
(b) Find the time t_1 needed to reach the highest point.
(c) Find the height H.
The particle falls back to the earth from the height H.
(d) Set up the equation of motion.
(e) Find the time t_2 needed to reach the ground.
(f) Show that the final velocity is equal to

$$v_f = \frac{v_0}{\sqrt{1 + \left(\dfrac{v_0}{\beta}\right)^2}}, \qquad \beta^2 = \frac{mg}{b}$$

(This problem is interesting insofar as a variety of integrals and mathematical identities is needed.)

4.12 Describe qualitatively the motion of a particle under the influence of the force $\mathbf{F} = -k\mathbf{r} \times \dot{\mathbf{r}}$. Do not attempt to solve the equations of motion.

4.13 Apply the energy integral (4.57) to derive the equation $y = -(1/2)gt^2 + v_0 t + y_0$ for the free fall.

4.14 The velocity of a particle of mass m varies with the distance x according to the equation $v = bx^{-n}$.

(a) Find the force acting on the particle as a function of x.
(b) Find the position as a function of the time. Set $x(0) = x_0$.
(c) Find the force as a function of the time.

4.15 Show that the equation of motion is integrable for the force $F(x, t) = f(x)g(t)$, where f and g are known functions. Set up a procedure for finding $x = x(t)$.

4.16 Solve the equations of motion for the forces and initial values given.

(a) $F(x, \dot{x}) = -a\dfrac{\dot{x}^2}{x}$, $x(0) = x_0$, $v(0) = v_0$

(b) $F(x, \dot{x}) = bx\dot{x}$, $x(0) = x_0$, $v(0) = v_0$

(c) $F(x, \dot{x}, t) = b(\dot{x}t + x)$, $x(0) = x_0$, $v(0) = 0$

chapter

5 Oscillations

No one knows what he can do until he tries.

LATIN PROVERB

5.1 INTRODUCTION

Oscillation of a particle is one of the most common forms of motion. It occurs when the particle is displaced from the stable equilibrium position x_0, the point of minimum potential energy (see Figure 5.1). For all points $x \neq x_0$ in the neighborhood of x_0, the force on the mass point is directed toward x_0. The points x_1 and x_2 of maximal displacement are determined by the value of the total energy E of the system:

$$V(x_1) = V(x_2) = E \qquad (5.1)$$

The time needed for the particle to traverse a complete cycle, that is, to move from x_1 to x_2 and back to x_1, is called the period T of oscillation. Its reciprocal

$$f = \frac{1}{T} \qquad (5.2)$$

is called the frequency of oscillation. It represents the number of oscillations per unit time.

5.2 THE LINEAR HARMONIC OSCILLATOR

5.2.1 Definition and Examples

The (periodic) motion of a particle is called harmonic if the position $x(t)$ can be described by a sine function. It will be shown below that a system undergoes harmonic motion if the potential energy function is of the form

$$V(x) = \tfrac{1}{2}kx^2 \qquad (5.3)$$

where k is a positive constant. A system that has a potential energy function of this form is called a linear harmonic oscillator, LHO for short. The potential energy function shown in Figure 5.1 does not represent an LHO, and the motion

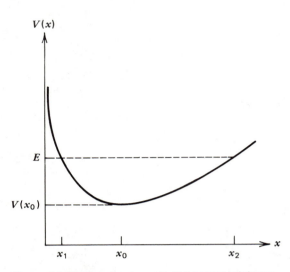

Figure 5.1 Motion about a potential energy minimum.

of the particle is not harmonic (see Figure 5.2). The force associated with the potential energy (5.3) is

$$F(x) = -\frac{dV}{dx} = -\frac{d}{dx}\left(\frac{1}{2}kx^2\right) = -kx \tag{5.4}$$

and the equation of motion for the LHO becomes

$$m\ddot{x} = F(x) = -kx$$

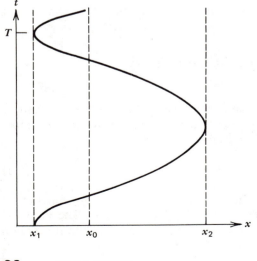

Figure 5.2 Time versus position graph of a particle subjected to the potential energy function drawn in Figure 5.1.

or

$$\ddot{x} + \frac{k}{m}x = 0 \qquad (5.5)$$

There are several important realizations of the LHO. In the following examples, the variable x is not always a position coordinate and the quantities k and m may be replaced by other constants.

The Mass-Spring System. A mass point attached to a spring performs harmonic oscillations about the equilibrium position $x_0 = 0$. Experiments show that the applied force on a spring (or any elastic object) produces an elastic deformation (compression or elongation) $\Delta x = x - x_0 = x$, which is directly proportional to the applied force:

$$F_{\text{appl}} = kx \qquad (5.6)$$

The constant k is called the spring constant. The force exerted by the spring on the mass point is the restoring force

$$F = -kx \qquad (5.7)$$

which is equal to the LHO force (Eq. 5.4). Equation 5.6 is known as Hooke's law.

The Torsion Pendulum. The torsion pendulum consists of an elastic rod (or wire) connected to a massive object with moment of inertia I. A torque N_{appl} applied to the rod produces an elastic angular deformation, measured in terms of the angle of twist θ, which is proportional to the applied torque:

$$N_{\text{appl}} = K\theta \qquad (5.8)$$

The constant K is called the torsion constant. The torque exerted by the bar on the object is the restoring torque

$$N = -K\theta \qquad (5.9)$$

Inserting this expression of the torque into (4.28) we obtain

$$I\ddot{\theta} = N = -K\theta$$

or

$$\ddot{\theta} + \frac{K}{I}\theta = 0 \qquad (5.10)$$

This differential equation has the same mathematical form as the LHO equation (5.5), hence, the angular oscillations must be harmonic.

The L-C Circuit. The electric charge stored on a capacitor of capacitance C, which is connected to an inductor of self-inductance L, satisfies the differential equation

$$\ddot{q} + \frac{1}{LC}q = 0 \tag{5.11}$$

Because this equation is of the form of Eq. 5.5, the electric charge on C varies harmonically.

Small Oscillations. The systems described in the examples above oscillate harmonically for any finite value of x, θ, or q, respectively, and the potential energy V expressed in terms of these variables is of the form of Eq. 5.3 for all these quantities. The potential energy $V(x)$ of an arbitrary oscillator, for example, the potential energy $V(x)$ depicted in Figure 5.1, may be approximated in the neighborhood of the point x_0 by a potential energy function of the form (5.3). This can be shown as follows: We expand the function $V(x)$ into a Taylor series about the equilibrium point x_0.

$$V(x) = V(x_0) + \frac{1}{1!}\frac{dV}{dx}\bigg|_{x=x_0}(x-x_0) + \frac{1}{2!}\frac{d^2V}{dx^2}\bigg|_{x=x_0}(x-x_0)^2$$

$$+ \frac{1}{3!}\frac{d^3V}{dx^3}\bigg|_{x=x_0}(x-x_0)^3 + \cdots \tag{5.12}$$

The first term on the right-hand side, $V(x_0)$, is the value of the potential energy at the equilibrium point. It may be set equal to zero because, as we have seen in the previous chapter, constants added to the potential energy have no influence on the motion of a particle. The second term vanishes identically because the force on the particle at the equilibrium point is zero:

$$F(x_0) = -\frac{dV}{dx}\bigg|_{x=x_0} = 0 \tag{5.13}$$

If the second and third derivatives of V at x_0 are of the same order of magnitude, the fourth term in (5.12) is small compared with the third for small displacements $\Delta x = (x - x_0)$ and can therefore be neglected. Under similar assumptions all other higher order terms may be omitted. Placing the equilibrium point at the origin, $x_0 = 0$, we finally obtain for small values of x

$$V_{\text{LHO}}(x) = \tfrac{1}{2}kx^2 \tag{5.14}$$

where

$$k = \frac{d^2V}{dx^2}\bigg|_{x=x_0} \tag{5.15}$$

Example 5.1

Find the LHO approximation for the potential energy function

$$V(x) = ax^{-2} + bx^2$$

Using (5.13) we first calculate the equilibrium position x_0.

$$\frac{dV}{dx} = -2ax^{-3} + 2bx = 0 \qquad \text{implies } x_0 = \left(\frac{a}{b}\right)^{1/4}$$

From (5.15) we obtain

$$k = \frac{d^2V}{dx^2}\Big|_{x=x_0} = (6ax^{-4} + 2b)\big|_{x=x_0} = 6a\frac{b}{a} + 2b = 8b$$

Therefore,

$$V_{\mathrm{LHO}}(x) = \tfrac{1}{2} \cdot 8bx^2 = 4bx^2 \qquad\blacksquare$$

The four typical realizations of the LHO are compared with each other in Table 5.1.

5.2.2 Solution of the Equation of Motion

We now proceed to solve the equation of motion for the LHO and, because of its great importance in physics, we shall apply two different methods.

The Energy Integral. Because the potential energy $V(x) = (\tfrac{1}{2})kx^2$ is a symmetric function of x, that is, $V(x) = V(-x)$, the motion of the particle must be symmetric relative to the point of equilibrium. The points of maximal displacement have the same magnitude $A : x_1 = -A, x_2 = +A$. The constant A is called the amplitude of the oscillation. The total energy E is equal to the potential energy value $V(\pm A)$ because the kinetic energy vanishes at the turning points $\pm A$:

$$E = T + V = V(\pm A) = \tfrac{1}{2}kA^2 \qquad (5.16)$$

Table 5.1   Realizations of the Linear Harmonic Oscillator

System	Variable	Inertial Factor	Force Factor	Angular Frequency
Mass-spring	x	m	k	$\sqrt{\dfrac{k}{m}}$
Torsion pendulum	θ	I	K	$\sqrt{\dfrac{K}{I}}$
LC-circuit	q	L	C^{-1}	$\sqrt{\dfrac{1}{LC}}$
Small oscillations	x	m	$k = \dfrac{d^2V(x)}{dx^2}\Big|_{x=x_0}$	$\sqrt{\dfrac{k}{m}}$

The velocity v as a function of the position is the given by

$$v = \pm \sqrt{\frac{2T}{m}} = \pm \sqrt{\frac{2}{m}[E - V(x)]} = \pm \sqrt{\frac{2}{m}[V(A) - V(x)]}$$

$$= \pm \sqrt{\frac{k}{m}} \sqrt{A^2 - x^2} \qquad (5.17)$$

The integral (4.57) is easily evaluated. We find that

$$t = \int_{x_0}^{x} \frac{dx'}{\sqrt{\frac{k}{m}(A^2 - x'^2)}} = \frac{1}{\sqrt{\frac{k}{m}}}\left(\text{arc sin}\frac{x}{A} - \text{arc sin}\frac{x_0}{A}\right)$$

$$= \frac{1}{\omega_0}\left(\text{arc sin}\frac{x}{A} - \delta\right) \qquad (5.18)$$

where we have introduced the two constants

$$\omega_0 = \sqrt{\frac{k}{m}} \qquad (5.19)$$

and

$$\delta = \text{arc sin}\frac{x_0}{A} \qquad (5.20)$$

Solving Eq. 5.18 for $x(t)$ we obtain

$$x(t) = A\sin(\omega_0 t + \delta) \qquad (5.21)$$

which proves that the motion is harmonic. The two constants ω_0 and δ are easily interpreted. When the time t has increased by T (the period of oscillation) the argument of the sine function has increased by the angle 2π. Because $x(t) = x(t + T)$, we obtain from Eq. 5.21 the result

$$\omega_0(t + T) + \delta = \omega_0 t + \delta + 2\pi$$

or

$$\omega_0 = \frac{2\pi}{T} = 2\pi f, \qquad (5.22)$$

which is the angular frequency of the oscillation. The angle δ, called the initial phase angle, determines the initial values of the position and of the velocity. It follows directly from Eq. 5.20 that

$$x(0) = x_0 = A\sin\delta$$

Differentiating (5.21) with respect to t we obtain

$$v(t) = A\omega_0 \cos(\omega_0 t + \delta),\qquad(5.23)$$

hence,

$$v(0) = v_0 = A\omega_0 \cos\delta\qquad(5.24)$$

Note that both x_0 and v_0 are needed to uniquely determine δ in the interval $[0, 2\pi]$ because the arc sine and arc cosine functions are multivalued functions. Figure 5.3 shows the relationships between x, v, and δ.

The Purely Mathematical Approach. The equation of motion (5.5) is a differential equation with the following properties:

1. It is of second order. The highest derivative is of order two.
2. It is homogeneous. An additive function $g(t)$ does not appear.
3. It is linear. If $x_1(t)$ and $x_2(t)$ are two solutions, then $ax_1(t) + bx_2(t)$, where a and b are arbitrary constants, is also a solution of the same differential equation.
4. It has constant coefficients: The factors multiplying x, $\dot{x}$, $\ddot{x}$, and so on, are constants.

For all homogeneous linear differential equations with constant coefficients the method of a trial function is convenient. One assumes that $x(t)$ is of the form

$$x(t) = Ae^{qt},\qquad(5.25)$$

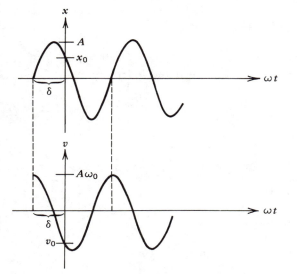

Figure 5.3 Harmonic oscillation. Position and velocity versus time.

which reduces the differential equation to an algebraic equation for q. Differentiating (5.25) twice with respect to t we obtain

$$\dot{x} = Aqe^{qt} = qx \qquad (5.26)$$

and

$$\ddot{x} = Aq^2e^{qt} = q^2x \qquad (5.27)$$

Combining (5.5) and (5.27), we find

$$q^2x + \frac{k}{m}x = 0$$

or, because $x(t) \neq 0$ in general,

$$q^2 = -\frac{k}{m} = -\omega_0^2$$

The last equation has the two solutions

$$q_+ = i\omega_0 \qquad \text{and} \qquad q_- = -i\omega_0 \qquad (5.28)$$

Hence the general solution $x(t)$ is

$$x(t) = A_+e^{i\omega_0 t} + A_-e^{-i\omega_0 t} \qquad (5.29)$$

Because $x(t)$ is real valued, the right-hand side must be real also which is only possible if A_+ and A_- are complex numbers. Comparing

$$x(t) = x^*(t) = A_+^* e^{-i\omega_0 t} + A_-^* e^{+i\omega_0 t}$$

with (5.29) we conclude that $A_- = A_+^*$. In order to put Eq. 5.29 into a more familiar form, we introduce the two real constants a and b defined by

$$A_+ = \tfrac{1}{2}(a - ib) \qquad \text{and} \qquad A_- = \tfrac{1}{2}(a + ib)$$

and apply Euler's formula (2.54) to (5.29):

$$x(t) = \tfrac{1}{2}(a - ib)(\cos \omega_0 t + i\sin \omega_0 t) + \tfrac{1}{2}(a + ib)(\cos \omega_0 t - i\sin \omega_0 t)$$
$$= a\cos \omega_0 t + b\sin \omega_0 t \qquad (5.30)$$

The connection with (5.21) is established by setting

$$a = A\sin \delta \qquad \text{and} \qquad b = A\cos \delta$$

Each of the three (equivalent) solutions (5.29), (5.30), and (5.21) is useful. Equation 5.29 has the simplest mathematical form, Eq. 5.30 exhibits most clearly the fact that the motion of the LHO is a superposition of a sine and a cosine function, and Eq. 5.21 relates to the initial values of x and v directly.

5.3 THE DAMPED HARMONIC OSCILLATOR

In real situations, oscillating systems are also acted upon by frictional forces that cause the total energy and the amplitude to decrease with time. We assume that the frictional force F_f is velocity dependent and of the simplest possible form

$$F_f = -b\dot{x}$$

where b is a constant. The equation of motion for the damped harmonic oscillator becomes

$$m\ddot{x} = F_{LHO} + F_f = -kx - b\dot{x}$$

which can be written in the form

$$\ddot{x} + 2\gamma\dot{x} + \omega_0^2 x = 0, \tag{5.31}$$

where ω_0 is the angular frequency (5.19) of the LHO and γ is defined by

$$\gamma = \frac{b}{2m} \tag{5.32}$$

The differential equation (5.31) is also linear and homogeneous, and it contains constant coefficients. Therefore, the method of the trial function may be reapplied to solve the equation. Substitution of (5.25), (5.26), and (5.27) into (5.31) gives

$$q^2 x + 2\gamma q x + \omega_0^2 x = 0$$

which after elimination of x becomes

$$q^2 + 2\gamma q + \omega_0^2 = 0$$

This algebraic equation has the solutions

$$q_+ = -\gamma + \sqrt{\gamma^2 - \omega_0^2} \quad \text{and} \quad q_- = -\gamma - \sqrt{\gamma^2 - \omega_0^2} \tag{5.33}$$

There are three types of the motion according to whether the radicand $\gamma^2 - \omega_0^2$ is positive, zero, or negative.

Overdamping. $\gamma^2 > \omega_0^2$. In this case both q_+ and q_- are negative, and the displacement

$$x_{od}(t) = A \exp\left[\left(-\gamma + \sqrt{\gamma^2 - \omega_0^2}\right)t\right] + B \exp\left[\left(-\gamma - \sqrt{\gamma^2 - \omega_0^2}\right)t\right] \tag{5.34}$$

approaches zero as t goes to infinity. The first term dominates over the second term for large values of t because $q_+ > q_-$. The motion is represented in Figure 5.4, where we have set $A = B = 1$ for simplicity.

Critical Damping. $\gamma^2 = \omega_0^2$. Because in this case $q_+ = q_- = -\gamma$, the method of the trial function yields only one solution:

$$x_1(t) = Ae^{-\gamma t} \tag{5.35}$$

Another solution must exist because the equation of the motion is of second order. For linear homogeneous differential equations the following procedure leads to the second solution. One inserts another trial function of the form

$$x_2(t) = f(t)x_1(t)$$

into the given differential equation, which yields a differential equation for the unknown function $f(t)$ of order one less than that of the original equation. Differentiating x_2 twice, we obtain

$$\dot{x}_2 = \dot{f}x_1 + f\dot{x}_1$$

and

$$\ddot{x}_2 = \ddot{f}x_1 + 2\dot{f}\dot{x}_1 + f\ddot{x}_1$$

If we insert x_2 and its derivatives into (5.32), we arrive at

$$\ddot{f}x_1 + 2\dot{f}\dot{x}_1 + f\ddot{x}_1 + 2\gamma\left(\dot{f}x_1 + f\dot{x}_1\right) + \omega_0^2 f x_1 = 0$$

The underlined terms vanish because x_1 is a solution of the differential equation (5.31). The dotted terms cancel each other because $\dot{x}_1 = -\gamma x_1$ by (5.35). It follows that the function f obeys the simple equation

$$\ddot{f} = 0$$

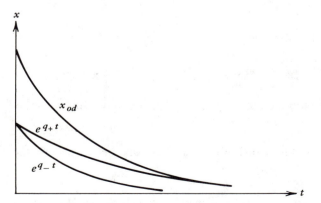

Figure 5.4 Overdamping.

which has $f = t$ as a solution. The second linearly independent solution of (5.32) is therefore,

$$x_2(t) = Bte^{-\gamma t}$$

such that the general solution becomes

$$x_{\text{crit}}(t) = Ae^{-\gamma t} + Bte^{-\gamma t} = e^{-\gamma t}(A + Bt) \tag{5.36}$$

It is instructive to compare equations 5.34 and 5.36 with each other. Because

$$\lim_{t \to \infty} \frac{x_{\text{crit}}}{x_{od}} = \lim_{t \to \infty} \frac{e^{-\gamma t}(A + Bt)}{e^{-\gamma t}\left(A'e^{\sqrt{\gamma^2 - \omega_0^2}\, t} + B'e^{-\sqrt{\gamma^2 - \omega_0^2}\, t}\right)} = 0,$$

the critical solution x_{crit} approaches zero faster than x_{od}. Use of critical damping is made in various measuring devices to achieve the fastest possible reading of the measured value of a physical quantity.

Underdamping. $\gamma^2 < \omega_0^2$. If we introduce the real constant

$$\omega = \sqrt{\omega_0^2 - \gamma^2}, \tag{5.37}$$

the parameters q_+ and q_- can be written

$$q_+ = -\gamma + i\omega \qquad \text{and} \qquad q_- = -\gamma - i\omega$$

and the general solution is

$$\begin{aligned}
x_{ud}(t) &= A_+ e^{(-\gamma + i\omega)t} + A_- e^{(-\gamma - i\omega)t} \\
&= e^{-\gamma t}\left(A_+ e^{i\omega t} + A_- e^{-i\omega t}\right) \\
&= Ae^{-\gamma t}\sin(\omega t + \delta)
\end{aligned} \tag{5.38}$$

The system performs harmonic oscillations, but the amplitude decreases exponentially (see Figure 5.5). The frequency of oscillation of the damped oscillator is

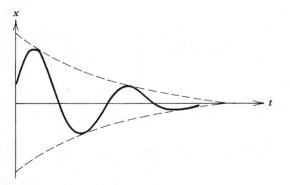

Figure 5.5 Underdamping.

smaller than that of the undamped oscillator by Eq. 5.37, but for very small damping the frequency remains practically unchanged: $\omega \approx \omega_0 - \gamma^2/2\omega_0$.

5.4   THE FORCED HARMONIC OSCILLATOR

In the previous section we investigated the oscillation of a mass point, once set in motion, under the influence of forces originating in the system itself. Often external time-dependent forces also act on the oscillator. In this case the equation of motion becomes

$$m\ddot{x} + b\dot{x} + kx = F(t) \qquad (5.39)$$

Because this differential equation is inhomogeneous, its general solution is the sum of the general solution of the associated homogeneous equation and of any particular solution of the inhomogeneous equation:

$$x_{gi} = x_{gh} + x_{pi}$$

The general solution of the homogeneous equation (5.31) was obtained in the preceding section, and our remaining task is to find any particular solution of (5.39). Various mathematical methods exist for finding such a solution in a closed form, but we shall not pursue the general problem here. Instead, we shall restrict ourselves to the important case of a periodic driving force with the frequency ω',

$$F(t) = F_0 \cos \omega' t.$$

With this force, the particular solution x_{pi} of (5.39) must also be periodic with the same frequency ω':

$$x_{pi} = A \cos \omega' t \qquad (5.40)$$

In general, the two quantities F and x will be out of phase by a relative phase angle φ, which we include in the force term. Then the equation of motion[1] reads

$$m\ddot{x} + b\dot{x} + kx = F_0 \cos(\omega' t + \varphi), \qquad (5.41)$$

or in terms of the constants ω_0, Eq. 5.19, and γ, Eq. 5.32,

$$\ddot{x} + 2\gamma\dot{x} + \omega_0^2 x = \frac{F_0}{m} \cos(\omega' t + \varphi) \qquad (5.42)$$

Inserting (5.40) into (5.42), we obtain

$$-A\omega'^2 \cos \omega' t - A2\gamma\omega' \sin \omega' t + A\omega_0^2 \cos \omega' t$$

$$= \frac{F_0}{m}(\cos \omega' t \cos \varphi - \sin \omega' t \sin \varphi)$$

---

[1] Equations of the form (5.41) appear for example in the theory of light dispersion.

Because the sine and cosine functions are linearly independent, the coefficients of $\sin \omega't$ and $\cos \omega't$, respectively, must be equal to each other:

$$A2\gamma\omega' = \frac{F_0}{m} \sin \varphi \qquad (5.43)$$

$$A\left(\omega_0^2 - \omega'^2\right) = \frac{F_0}{m} \cos \varphi \qquad (5.44)$$

Dividing (5.44) into (5.43), we obtain for the phase angle the equation

$$\tan \varphi = \frac{2\gamma\omega'}{\omega_0^2 - \omega'^2} \qquad (5.45)$$

The amplitude A is found by taking the square root of the sum of the squares of (5.43) and (5.44):

$$A = \frac{F_0/m}{\left[\left(\omega_0^2 - \omega'^2\right)^2 + 4\gamma^2\omega'^2\right]^{1/2}} \qquad (5.46)$$

The amplitude has a maximum when the driving frequency takes on the value

$$\omega_r = \sqrt{\omega_0^2 - 2\gamma^2} \qquad (5.47)$$

which is called the resonance frequency. The oscillator is then said to be in resonance with the periodic driving force. Observe the difference between ω, Eq. 5.37, and ω_r, Eq. 5.47. For small damping we obtain $\omega_r \approx \omega_0 - \gamma^2/\omega_0$, while $\omega \approx \omega_0 - \gamma^2/2\omega_0$. In the case of resonance the amplitude becomes

$$A_r = \frac{F_0/m}{2\gamma\sqrt{\omega_0^2 - \gamma^2}} = \frac{F_0}{2m\gamma\omega}$$

As the damping factor γ goes to zero, the resonance amplitude increases indefinitely. Resonance may cause severe damage to an oscillating system if it is only weakly damped.

The phase relation between F and x is also of interest.

1. If $\omega' < \omega_0$, then $0 \leqslant \varphi < \pi/2$, and F oscillates ahead of x by the angle φ.
2. If $\omega' = \omega_r$, then $\tan \varphi = \omega_r^2/\gamma$, and $\varphi < \pi/2$ for resonance.
3. If $\omega' = \omega_0 > \omega_r$, then $\varphi = \pi/2$, and maximum force is applied when the system passes through the equilibrium point.
4. If $\omega' \gg \omega_0$, then $\varphi \lesssim \pi$, and the motion of the mass point is opposite to the applied force.

Inspecting the general solution of (5.41),

$$x(t) = A_+ e^{q_+ t} + A_- e^{q_- t} + A \cos \omega't, \qquad (5.48)$$

we see that

$$x(t) \approx A \cos \omega' t$$

for large values of t. That part of (5.48) that arises from the associated homogeneous differential equation and that vanishes as t approaches infinity is therefore called the transient term. Although A is given by (5.46), the integration constants A_+ and A_- are found from the initial values x_0 and v_0 of the system.

5.5 THE HARMONIC OSCILLATOR IN TWO AND THREE DIMENSIONS

Let us consider now a mass point that is free to undergo harmonic oscillations in three dimensions. Choosing an appropriate coordinate system, we can write the three-dimensional restoring force in the general form

$$\mathbf{F} = -(k_1 x \mathbf{i} + k_2 y \mathbf{j} + k_3 z \mathbf{k}) \tag{5.49}$$

Because the force $\mathbf{F}$ is separable, the vector equation of motion splits into three decoupled differential equations of the LHO type (5.5), which have the solutions

$$x(t) = A_1 \sin(\omega_1 t + \delta_1), \tag{5.50a}$$

$$y(t) = A_2 \sin(\omega_2 t + \delta_2), \tag{5.50b}$$

$$z(t) = A_3 \sin(\omega_3 t + \delta_3) \tag{5.50c}$$

where the angular frequencies ω_i are given by

$$\omega_i = \sqrt{\frac{k_i}{m}} \quad i = 1, 2, 3 \tag{5.51}$$

The particle is confined to move inside a rectangular box of side lengths $2A_1$, $2A_2$, and $2A_3$.

The orbit of the particle can be visualized more easily when we restrict ourselves to the case of the motion in two dimensions. Without loss of generality, we choose $x(0) = x_0 = 0$ as the initial value for the x-coordinate, which implies that $\delta_1 = 0$. Then $\varphi = \delta_2$ is the relative phase angle between $x(t)$ and $y(t)$. Solving (5.50a) for t and inserting this value into (5.50b), we obtain the orbit equation

$$y = A_2 \sin\left(\frac{\omega_2}{\omega_1} \arcsin \frac{x}{A_1} + \varphi\right) \tag{5.52}$$

The orbit is closed only if ω_2/ω_1 is a rational number, that is, is equal to N_2/N_1 where N_1 and N_2 are integers. Because

$$\frac{N_2}{N_1} = \frac{\omega_2}{\omega_1} = \frac{f_2}{f_1} = \frac{T_1}{T_2},$$

we find that after the elapsed time

$$t = N_1 T_1 = N_2 T_2 \tag{5.53}$$

the oscillator has undergone N_1 oscillations in the x-direction and N_2 oscillations in the y-direction. If ω_2/ω_1 is an irrational number, the orbit is not closed, and the mass point comes close at some time to any point within the rectangle defined by the amplitudes A_1 and A_2. The patterns of the orbits are called Lissajous figures, and many have great aesthetic appeal. They can be displayed most easily on an oscilloscope. Some orbits are shown in Figure 5.6.

A system for which the restoring force is given by

$$\mathbf{F} = -k\mathbf{r},$$

is called an isotropic harmonic oscillator. Because $k = k_1 = k_2 = k_3$, the mass point oscillates in all directions with the same frequency

$$\omega_1 = \omega_2 = \omega_3 = \omega = \sqrt{\frac{k}{m}} \tag{5.54}$$

Considering again two-dimensional motion, and setting $\delta_1 = 0$ and $\varphi = \delta_2$, we

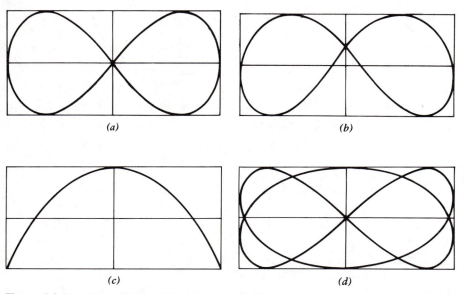

(a)

(b)

(c)

(d)

Figure 5.6 Lissajous' figures: (a) $\omega_1 : \omega_2 = 1 : 2$; (b) $\omega_1 : \omega_2 = 1 : 2$; (c) $\omega_1 : \omega_2 = 1 : 2$; (d) $\omega_1 : \omega_2 = 2 : 3$.

obtain from (5.50) and (5.54)

$$y = A_2 \sin\left(\arcsin\frac{x}{A_1} + \varphi\right)$$

$$= A_2 \left[\sin\left(\arcsin\frac{x}{A_1}\right)\cos\varphi + \cos\left(\arcsin\frac{x}{A_1}\right)\sin\varphi\right]$$

$$= \frac{A_2}{A_1}\left(x\cos\varphi + \sqrt{A_1^2 - x^2}\,\sin\varphi\right)$$

The phase angle φ determines the shape of the possible orbits. If $\varphi = 0, \pi$, then $y = \pm(A_2/A_1)x$, and the orbits are straight lines. If $\varphi = \pi/2, 3\pi/2$, then $y = \pm(A_2/A_1)\sqrt{A_1^2 - x^2}$ or $(x/A_1)^2 + (y/A_2)^2 = 1$, and the orbit is an ellipse. For $\varphi = \pi/2$ the particle moves clockwise, for $\varphi = 3\pi/2$ it moves counterclockwise. For all other values of the phase angle, the orbit is an ellipse inside a rectangle of side lengths $2A_1$ and $2A_2$. The ellipse is rotated relative to the Cartesian

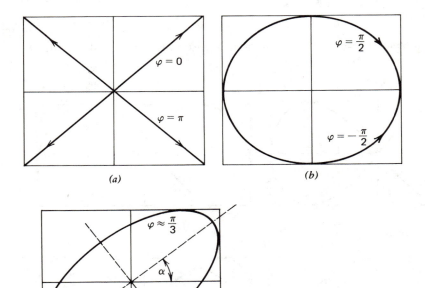

(a)　　　　　　　　　(b)

(c)

Figure 5.7 Orbits of a two-dimensional isotropic harmonic oscillator for various phase angles.

coordinates by some angle α. A tedious calculation yields

$$\tan 2\alpha = \frac{2\cos\varphi}{C} \quad \text{where} \quad C = \frac{A_1}{A_2} - \frac{A_2}{A_1}$$

Some orbits are shown in Figure 5.7.

5.6 THE SIMPLE PENDULUM

A system that oscillates as a whole about a fixed point under the influence of the gravitational force (weight), is called a pendulum. It is called a physical pendulum if the mass is distributed over the entire system. It is called a mathematical or simple pendulum if the system can be thought of as a mass point attached to an assumed massless cord or rod. The mass point may move on the surface of a sphere (spherical pendulum) or it may be restricted to oscillate in a plane. In the first case the orbit will have a complicated form depending on the initial conditions, whereas in the latter case the orbit is simply a portion of a circle. We shall discuss the motion of the simple pendulum in a plane only, but we shall use two different approaches.

Small Oscillations. We choose the x-coordinate as the parameter of the circular orbit of radius l and take $x_0 = 0$ as the equilibrium point Q (see Figure 5.8). The potential energy of the mass point at P relative to Q is equal to

$$V(x) = mgh = mg\left(l - \sqrt{l^2 - x^2}\right) \tag{5.55}$$

This potential energy function is clearly not of the form $(\frac{1}{2})kx^2$ of Eq. 5.3, and therefore the motion will not be harmonic. Also, the integral (Eq. 4.57)

$$t = \int_{x_0}^{x} \frac{dx}{\sqrt{\frac{2}{m}\left[E - mg\left(l - \sqrt{l^2 - x^2}\right)\right]}}$$

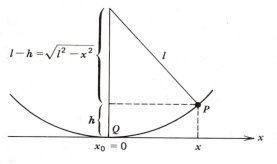

$l - h = \sqrt{l^2 - x^2}$

Figure 5.8 The simple pendulum, Cartesian coordinates.

cannot be evaluated in terms of elementary functions. In order to find the frequency of oscillation, we calculate the LHO approximation of the potential energy (5.55):

$$\frac{dV}{dx} = mg\left[-\frac{1}{2}\frac{1}{\sqrt{l^2 - x^2}}(-2x) \right] = mg\frac{x}{\sqrt{l^2 - x^2}},$$

$$k = \frac{d^2V}{dx^2}\bigg|_{x=x_0=0} = mg\left(\frac{1}{\sqrt{l^2 - x^2}} + x\frac{d}{dx}\frac{1}{\sqrt{l^2 - x^2}} \right)\bigg|_{x=0}$$

$$= \frac{mg}{l} \tag{5.56}$$

Hence the potential energy V_{LHO} of (5.55) is equal to

$$V_{\text{LHO}} = \frac{1}{2}\frac{mg}{l}x^2,$$

and the frequency of oscillation becomes

$$\omega = \sqrt{\frac{k}{m}} = \sqrt{\frac{mg}{l}\cdot\frac{1}{m}} = \sqrt{\frac{g}{l}} \tag{5.57}$$

for small amplitudes. For the period T we obtain the well-known formula

$$T = 2\pi\sqrt{\frac{l}{g}}$$

The Equation of Motion.   We choose θ as the parameter of the circle and apply Eq. 4.28 for circular motion. The moment of inertia is $I = ml^2$. The restoring torque becomes

$$-|\mathbf{N}| = -lmg\sin\theta,$$

(see Figure 5.9). We then obtain from (4.28)

$$I\ddot{\theta} = ml^2\ddot{\theta} = -N = -lmg\sin\theta$$

or

$$\ddot{\theta} + \frac{g}{l}\sin\theta = 0 \tag{5.58}$$

which for small angles ($\sin\theta \approx \theta$) reduces to the LHO form

$$\ddot{\theta} + \frac{g}{l}\theta = 0$$

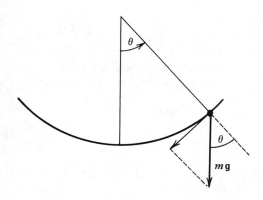

Figure 5.9 The simple pendulum, polar coordinates.

The solution of the differential equation (5.58) leads to elliptic integrals that can be evaluated by numerical methods only. Although elliptical integrals are of great importance and appear in numerous problems in physics, we shall not discuss them here. It turns out as expected that the period of oscillation depends on the angular amplitude θ_0. The function $T(\theta_0)$ is shown in Figure 5.10. For $\theta = 160°$ the period of oscillation is about twice that of small oscillations. For small values of θ_0 (measured in radians) the period T is approximately equal to

$$T = 2\pi\sqrt{\frac{l}{g}}\left(1 + \frac{\theta_0^2}{16}\right)$$

5.7 ISOCHRONOUS OSCILLATIONS

It was shown in Section 5.2 that the period of oscillation of a linear harmonic oscillator is independent of the amplitude. Amplitude independent oscillations are called isochronous.

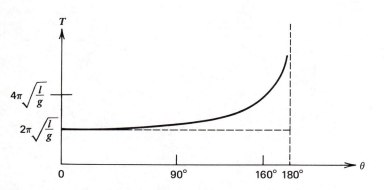

Figure 5.10 Period of oscillation as a function of the angular amplitude of a simple pendulum.

We consider now a mass point that is constrained to move on a curve and that may perform oscillations under the influence of an external force. We choose the arc length as the parameter of the curve measured from the point of equilibrium where $s = 0$. In order for the mass point to perform isochronous oscillations the tangential component of the applied force must be equal to the restoring force of the LHO, which is proportional to the displacement s. The general condition for isochronous oscillation is therefore,

$$F_{\text{tang}} = -ks \qquad (5.59)$$

We wish to find the equation of the curve if the external force is the constant gravitational force. In terms of the slope angle θ (see Figure 5.11), the tangential component of the weight is $F_{\text{tang}} = -mg \sin \theta$ which together with Eqs. 5.59 and 5.56 gives

$$s = \frac{mg}{k} \sin \theta = \frac{mg}{\dfrac{mg}{l}} \sin \theta = l \sin \theta$$

It is more convenient to express x and y as functions of θ. Using the identities

$$\frac{dx}{d\theta} = \frac{dx}{ds} \frac{ds}{d\theta} = \cos \theta \cdot l \cos \theta = l \cos^2 \theta$$

and

$$\frac{dy}{d\theta} = \frac{dy}{ds} \frac{ds}{d\theta} = \sin \theta \cdot l \cos \theta = l \sin \theta \cos \theta$$

we find

$$x(\theta) = \int_0^\theta l \cos^2 \theta \, d\theta = \frac{l}{4} (2\theta + \sin 2\theta)$$

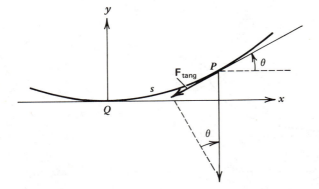

Figure 5.11 Relation between the weight, its tangential component, and the slope angle.

and

$$y(\theta) = \int_0^\theta l \sin \theta \cos \theta \, d\theta = \frac{l}{4}(1 - \cos 2\theta)$$

These are the equations of a cycloid in parameter form. They are related to the standard form $x = a(\varphi - \sin \varphi)$ and $y = a(1 - \cos \varphi)$ by translations in both the x and y directions and by a reflection of the y-axis.

The isochronous problem was first investigated by C. Huygens. He found the above solution and showed how an isochronous pendulum could be constructed by applying the fact that the evolute of a cycloid is also a cycloid (see Figure 5.12). The cycloid pendulum has never found widespread application, and electronic clocks have proven to be far superior and more accurate than any mechanical oscillating system.

5.8 MOTION OF A PARTICLE UNDER THE INFLUENCE OF A MAGNETIC FORCE

Although the magnetic force

$$\mathbf{F}_m = q(\mathbf{v} \times \mathbf{B}) \tag{5.60}$$

acting on an electrically charged particle of charge q which traverses a magnetic field is clearly not a harmonic oscillator force, the process of solving the equation of motion involving $\mathbf{F}_m$ may lead to equations of the LHO type as the following example demonstrates.

We consider a particle moving through a region of a constant magnetic field $\mathbf{B} = B\mathbf{k}$ pointing in the direction of the positive z-axis. We choose the initial values

$$\mathbf{r}_0 = (0, R, 0) \qquad \text{and} \qquad \mathbf{v}_0 = \left(v_{x_0}, 0, v_{z_0}\right)$$

The equation of motion

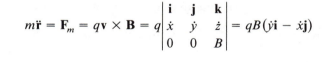

$$m\ddot{\mathbf{r}} = \mathbf{F}_m = q\mathbf{v} \times \mathbf{B} = q\begin{vmatrix} \mathbf{i} & \mathbf{j} & \mathbf{k} \\ \dot{x} & \dot{y} & \dot{z} \\ 0 & 0 & B \end{vmatrix} = qB(\dot{y}\mathbf{i} - \dot{x}\mathbf{j})$$

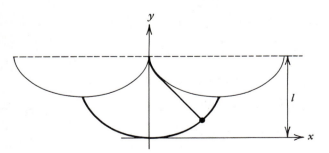

Figure 5.12 The evolute of a cycloid.

decomposes into the three equations

$$\ddot{x} - \omega_c \dot{y} = 0 \tag{5.61a}$$

$$\ddot{y} + \omega_c \dot{x} = 0 \tag{5.61b}$$

$$\ddot{z} = 0 \tag{5.61c}$$

where the constant ω_c is defined by

$$\omega_c = \frac{qB}{m} \tag{5.62}$$

Performing one integration of Eqs. 5.61a and 5.61b, we obtain

$$\dot{x} - \omega_c y = b' = \omega_c b, \tag{5.63a}$$

$$\dot{y} + \omega_c x = a' = \omega_c a \tag{5.63b}$$

where a and b are integration constants. Substituting $\dot{y}$ from (5.63b) into (5.61a), we find

$$\ddot{x} + \omega_c^2(x - a) = 0 \tag{5.64}$$

which has the solution

$$x(t) - a = A \sin \omega_c t + B \cos \omega_c t \tag{5.65a}$$

Similarly, substituting $\dot{x}$ from (5.63a) into (5.61b), we obtain

$$\ddot{y} + \omega_c^2(y - b) = 0$$

which has the solution

$$y(t) - b = C \sin \omega_c t + D \cos \omega_c t \tag{5.65b}$$

The values of the six integration constants a, b, A, B, C, and D are found as follows. Inserting $x(t)$ and $y(t)$ into (5.63a), we obtain

$$A\omega_c \cos \omega_c t - B\omega_c \sin \omega_c t - \omega_c(C \sin \omega_c t + D \cos \omega_c t - b) = \omega_c b$$

This equation must hold for all values of t, hence,

$$A - D = 0 \quad \text{and} \quad B + C = 0 \tag{5.66a}$$

Four more equations are obtained from the initial conditions

$$x(0) = 0 \text{ applied to } (5.65a): \ -a = B$$

$$y(0) = 0 \text{ applied to } (5.65b): \ R - b = D$$

$$v_x(0) = v_{x_0} \text{ applied to } (5.63a): \ v_{x_0} = \omega_c A \tag{5.66b}$$

$$v_y(0) = 0 \text{ applied to } (5.63b): \ 0 = C$$

110   OSCILLATIONS

The integration constants found from the set of (5.66) are

$$a = B = C = 0, \quad A = D = \frac{v_{x_0}}{\omega_c}, \quad b = R - \frac{v_{x_0}}{\omega_c}$$

Thus, for the x- and y-coordinates we obtain the result

$$x(t) = \frac{v_{x_0}}{\omega_c} \sin \omega_c t, \tag{5.67a}$$

$$y(t) = R - \frac{v_{x_0}}{\omega_c} + \frac{v_{x_0}}{\omega_c} \cos \omega_c t \tag{5.67b}$$

These are the equations of a circle of radius v_{x_0}/ω_c centered at the point $(0, R - v_{x_0}/\omega_c)$. If we choose R equal to the radius of the circle

$$R = \frac{v_{x_0}}{\omega_c} = \frac{m v_{x_0}}{qB}, \tag{5.68}$$

Eqs. 5.67a and 5.67b become

$$x(t) = R \sin \omega_c t$$

$$y(t) = R \cos \omega_c t$$

Because the solution of Eq. 5.61c is

$$z(t) = v_{z_0} t + z_0 = v_{z_0} t$$

which describes uniform motion in the z direction, the particle's orbit is a left-handed circular helix about the z-axis (see Figure 5.13).

The constant ω_c introduced in (5.62) is the angular frequency of revolution of the particle about the z-axis. The term ω_c is called the cyclotron frequency because the operation of cyclotrons rests upon the fact that ω_c is independent of the velocity of the particle and of the radius of the circle in nonrelativistic approximation.

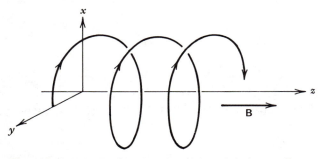

Figure 5.13 Helical orbit of a charged particle in a uniform magnetic field.

PROBLEMS

5.1 Find the effective spring constant for two springs of stiffness k_1 and k_2, respectively, if they are connected

(a) in series
(b) parallel

5.2 If a mass-spring system undergoes harmonic oscillations, the inertial mass m_s of the spring should be taken into account. Assuming that the spring has a uniform mass distribution and is uniformly stretched, show that the effective mass of the system is $m + m_s/3$.

5.3 A disc undergoes rotational motion in such a way that the rate of change of the angular speed is proportional to the angular displacement θ. Find θ, ω, and α as functions of the time if initially $\theta(0) = 0$ and $\omega(0) = \omega_0$.

5.4 The speed of a particle undergoing simple harmonic motion is suddenly doubled when the particle's displacement is one-half the amplitude. What is the new amplitude?

5.5 Find the LHO approximations and the frequencies of oscillation for particles of mass m oscillating about the points of equilibrium for the following potential energies:

(a) $V(x) = \dfrac{A}{x} + Bx$ (b) $V(x) = \dfrac{A}{x} + Bx^3$

(c) $V(x) = \dfrac{A}{x^2} - \dfrac{B}{x}$ (d) $V(x) = \dfrac{Ax^2}{x^2 + a^2}$

(e) $V(x) = A \sin bx$ (f) $V(x) = A \cosh bx - B \sinh bx \ (A > B)$

5.6 Is the method of small oscillations applicable to the potential energy function $V(x) = cx^4$? Explain.

5.7 A mass point is placed on a horizontal table on the surface of the earth where it performs small oscillations without friction. It is under the influence of the gravitational force described by the potential energy function $V(r) = -GMm/r$. Find the period of oscillation.

5.8 The time average of a function $f(t)$ is defined by

$$\bar{f} = \frac{1}{T} \int_0^T f(t)\, dt$$

Find the time averages of the kinetic and potential energies for the linear harmonic oscillator evaluated over one period.

5.9 Using the equation of motion, show that the total energy of the damped harmonic oscillator decreases with the time, that is, that $dE/dt < 0$.

5.10 Solve the equations of motion

(a) $m\ddot{x} - b\dot{x} + kx = 0$

(b)
$$m\ddot{x} + b\dot{x} - kx = 0$$

Discuss the possible cases of the motion.

5.11 Derive Eqs. 5.46 and 5.47.

5.12 Determine the integration constants A_+ and A_- in Eq. 5.48 for the forced harmonic oscillator if the initial values of the position and the velocity are:

(a)
$$x_0 \neq 0, \quad v_0 \neq 0$$

(b)
$$x_0 = 0, \quad v_0 = 0$$

In particular, consider the case of the underdamped oscillator.

5.13 A linear harmonic oscillator is acted upon by a time-dependent force of the form

$$F(t) = \left\{ \begin{matrix} F_0 = \text{constant}, & 0 \leqslant t \leqslant \tau \\ 0 & \tau < t \end{matrix} \right\}$$

The initial values are $x_0 = 0$ and $v_0 = 0$. Solve the equation of motion. Assume that both x and v are continuous at $t = \tau$. Investigate the special cases $\tau = T/2$ and $\tau = nT$ where n is an integer.

5.14 The potential energy of a nonisotropic two-dimensional harmonic oscillator is given by $V(x, y) = (\frac{1}{2})x^2 + (\frac{9}{2})y^2$. Solve the equations of motion for a particle with unit mass. Let $\mathbf{r}_0 = A_2 \mathbf{j}$ and $\mathbf{v}_0 = A_1 \mathbf{i}$. Draw the orbit.

5.15 By choosing an appropriate coordinate system and initial values the solution of the two-dimensional isotropic harmonic oscillator may be written in the form $x(t) = A \cos \omega_0 t$ and $y(t) = B \sin \omega_0 t$.

(a) Show that $l = (km)^{1/2}AB$ and $E = (\frac{1}{2})k(A^2 + B^2)$.
(b) Find A and B in terms of l and E. Assume that $A > B$.
(c) Derive the equation

$$r^2 = \frac{E}{k} \left[1 + \sqrt{1 - \left(\frac{l\omega_0}{E} \right)^2} \cos 2\omega_0 t \right]$$

chapter

6 Central Forces

It is a capital mistake to theorize before one has data.

SIR ARTHUR CONAN DOYLE

6.1 INTRODUCTION

It was shown in Chapter 4 that central forces

$$\mathbf{F}(\mathbf{r}) = f(\mathbf{r})\mathbf{e}_r \qquad (6.1)$$

leave the angular momentum of a particle unchanged, and furthermore that central isotropic forces

$$\mathbf{F}(\mathbf{r}) = f(r)\mathbf{e}_r \qquad (6.2)$$

are conservative, that is, leave the total mechanical energy of a particle constant. The conservation laws are consequences of the large degree of symmetry of the force, namely, radial symmetry. Central isotropic forces are of great importance because numerous forces of this kind exist in nature; for example:

(a) Gravitational force (Newton)

$$f(r) = -G\frac{m_1 m_2}{r^2}$$

(b) Electrostatic force (Coulomb)

$$f(r) = k_e \frac{q_1 q_2}{r^2}$$

(c) Intermolecular force (Van der Waals, Lennard-Jones)

$$f(r) = \frac{C_1}{r^{13}} - \frac{C_2}{r^7}$$

(d) Nuclear force (Yukawa)

$$f(r) = \left(\frac{C_1}{r} + \frac{C_2}{r^2} \right) e^{-C_3 r}$$

6.2 THE EQUATION OF MOTION

Because the angular momentum vector $\mathbf{l} = \mathbf{r} \times \mathbf{p} = m\mathbf{r} \times \mathbf{v}$ remains constant when the particle is acted upon by a central force, and because both $\mathbf{r}$ and $\mathbf{v}$ are perpendicular to $\mathbf{l}$, the orbit of the particle will always lie in a plane perpendicular to $\mathbf{l}$. It is therefore sufficient to treat central force problems in two dimensions only.

It would be unwise to set up the equations of motion in Cartesian coordinates because the force is dependent on x and y in a complicated way. The equations of motion

$$m\ddot{\mathbf{r}} = \mathbf{F} = f(\mathbf{r})\mathbf{e}_r \tag{6.3}$$

would in component form read

$$m\ddot{x} = f(x, y)\frac{x}{\sqrt{x^2 + y^2}}$$

and

$$m\ddot{y} = f(x, y)\frac{y}{\sqrt{x^2 + y^2}}$$

and we could hardly expect to find closed form solutions for x and y as functions of t from these equations.

Because the central force points in the radial direction, the choice of (plane) polar coordinates is much more natural. Applying (3.9d) to (6.3), and writing r instead of ρ, we obtain

$$m(\ddot{r} - r\dot{\varphi}^2) = f(r, \varphi) \tag{6.4}$$

and

$$m(2\dot{r}\dot{\varphi} + r\ddot{\varphi}) = 0 \tag{6.5}$$

These two equations are still coupled, but (6.5) immediately admits one integration. Equation 6.5 multiplied by r is simply the time derivative of the magnitude of the (conserved) angular momentum:

$$0 = m(2r\dot{r}\dot{\varphi} + r^2\ddot{\varphi}) = \frac{d}{dt}(mr^2\dot{\varphi}) = \frac{dl}{dt}$$

From

$$l = mr^2\dot{\varphi} = \text{constant} \tag{6.6}$$

follows

$$\dot{\varphi} = \frac{l}{mr^2} \tag{6.7}$$

which connects the two variables φ and r in a simple way.

For central isotropic forces, f is a function of r alone, and (6.4) after elimination of $\dot{\varphi}$ using (6.7), becomes

$$m\left(\ddot{r} - \frac{l^2}{m^2}\frac{1}{r^3}\right) = f(r) \tag{6.8}$$

This second-order differential equation can be solved for r as a function of t. Inserting $r(t)$ into (6.7) and integrating over t we find φ as a function of t: $\varphi = \varphi(t)$. The set $\{r(t), \varphi(t)\}$, which constitutes the solution of the equations of motion, describes the orbit of the particle in parametric form with the time t as the parameter.

In order to find the orbit equation in the form $r = r(\varphi)$, the following two procedures are possible.

One may solve the equation $\varphi = \varphi(t)$ for t and then substitute $t(\varphi)$ into the equation $r = r(t)$ such that $r = r(t(\varphi)) = r(\varphi)$. This procedure is somewhat cumbersome because two differential equations, namely (6.8) and then (6.7), need to be solved. It is simpler to set up a differential equation that relates r to φ directly. For this purpose we eliminate the time variable from the differential equation (6.8) using (6.7). First, we multiply Eq. 6.8 by $-(m/l^2)r^2$, which gives

$$-\left(\frac{m}{l}\right)^2 r^2 \ddot{r} + \frac{1}{r} = -\frac{m}{l^2}r^2 f(r) \tag{6.9}$$

It is convenient in most central force problems to use the variable

$$u = \frac{1}{r} \tag{6.10}$$

rather than r itself. The time derivative of $r(\varphi(t))$ becomes

$$\frac{dr}{dt} = \frac{d\left(\frac{1}{u}\right)}{dt} = -\frac{1}{u^2}\frac{du}{dt} = -\frac{1}{u^2}\frac{du}{d\varphi}\frac{d\varphi}{dt}$$

$$= -\frac{1}{u^2}\frac{du}{d\varphi}\frac{l}{m}u^2 = -\frac{l}{m}\frac{du}{d\varphi} \tag{6.11}$$

where we replaced $d\varphi/dt$ by $(l/m)u^2$ because of (6.7). The second derivative of r is

$$\frac{d^2r}{dt^2} = -\frac{l}{m}\frac{d}{dt}\left(\frac{du}{d\varphi}\right) = -\frac{l}{m}\frac{d}{d\varphi}\left(\frac{du}{d\varphi}\right)\frac{d\varphi}{dt} = -\left(\frac{l}{m}\right)^2 u^2 \frac{d^2u}{d\varphi^2}$$

Substituting the last expression for $\ddot{r}$ into (6.9), we find

$$-\left(\frac{m}{l}\right)^2 \frac{1}{u^2}(-1)\left(\frac{l}{m}\right)^2 u^2 \frac{d^2u}{d\varphi^2} + u = -\frac{m}{l^2}\frac{1}{u^2}f\left(\frac{1}{u}\right)$$

or, writing

$$u'' = \frac{d^2u}{d\varphi^2}$$

we arrive at

$$u'' + u = -\frac{m}{l^2}\frac{1}{u^2}f\left(\frac{1}{u}\right), \tag{6.12}$$

which is the desired differential equation for the orbit. Equation 6.12 may be employed to find the orbit $r(\varphi)$ if the force law is known, or to find the force law $f(r)$ if the equation $r = r(\varphi)$ for the orbit is known. Methods for integrating Eqs. 6.8 and 6.12 will be discussed in the next section.

Example 6.1

The orbit equation of a particle is $r = A\varphi^2$. What is the force law?
We differentiate $u = A^{-1}\varphi^{-2}$ with respect to φ:

$$u' = -\frac{2}{A}\varphi^{-3}, \quad u'' = \frac{6}{A}\varphi^{-4} = 6Au^2$$

Equation 6.12 becomes

$$f\left(\frac{1}{u}\right) = -\frac{l^2}{m}u^2(u'' + u) = -\frac{l^2}{m}u^2(6Au^2 + u)$$

or

$$f(r) = -\frac{l^2}{m}\left(\frac{6A}{r^4} + \frac{1}{r^3}\right) \qquad \blacksquare$$

Example 6.2

Find r as a function of t for the particle described in Example 6.1. Applying Eq. 6.7 to eliminate $\dot{\varphi}$, we obtain

$$\dot{r} = (A\varphi^2)^{\bullet} = 2A\varphi\dot{\varphi} = 2A\left(\frac{r}{A}\right)^{1/2}\frac{l}{mr^2} = \frac{2l}{m\sqrt{A}}r^{-3/2}$$

Evaluation of the integrals

$$\int_{r_0}^{r} r^{3/2}\,dr = \frac{2l}{m\sqrt{A}}\int_0^t dt$$

and rearrangement of the terms gives

$$r(t) = \left(\frac{5l}{m\sqrt{A}} t + r_0^{5/2} \right)^{2/5}$$

    ■

6.3 THE ENERGY INTEGRALS

We have seen in the preceding section that the theorem of angular momentum conservation follows from the angular part of the equations of motion Eq. 6.5. We now proceed to derive the theorem of energy conservation. For this purpose we multiply the radial part of the equations of motion Eq. 6.8, by $\dot{r}$:

$$m\ddot{r}\dot{r} - \frac{l^2}{m}\frac{\dot{r}}{r^3} = f(r)\dot{r} \tag{6.13}$$

The left side of (6.13) is equal to

$$\frac{d}{dt}\left(\frac{1}{2}m\dot{r}^2 + \frac{l^2}{2m}\frac{1}{r^2} \right)$$

whereas the right side is the time derivative of the potential energy

$$f(r)\dot{r} = -\frac{dV(r)}{dr}\frac{dr}{dt} = -\frac{dV}{dt}$$

Equation 6.13 becomes

$$\frac{d}{dt}\left\{ \frac{1}{2}m\dot{r}^2 + \frac{l^2}{2m}\frac{1}{r^2} + V(r) \right\} = 0$$

or

$$\frac{1}{2}m\dot{r}^2 + \frac{l^2}{2m}\frac{1}{r^2} + V(r) = \text{constant} = E \tag{6.14}$$

The first term is the radial contribution to the kinetic energy

$$T_{\text{rad}} = \frac{1}{2}m\dot{r}^2$$

whereas the second term is the kinetic energy due to angular motion:

$$T_{\text{ang}} = \frac{l^2}{2m}\frac{1}{r^2} = \frac{1}{2m}(mr^2\dot{\varphi})^2\frac{1}{r^2} = \frac{1}{2}mr^2\dot{\varphi}^2$$

The total energy (6.14) may be viewed in two ways. For physical reasons, one may combine the two contributions T_{rad} and T_{ang} into one term T, the kinetic energy of the particle. Obviously, this result follows faster from Eq. 3.9c. Then

$$E = T_{rad} + T_{ang} + V = T + V$$

as expected. For mathematical reasons one may combine T_{ang} (which, because of angular momentum conservation, depends on r only) with $V(r)$ to form the effective potential energy

$$V_{eff}(r) = T_{ang}(r) + V(r) \qquad (6.15)$$

The dependence of the total energy

$$E = T_{rad} + V_{eff}(r) = \tfrac{1}{2}m\dot{r}^2 + V_{eff}(r) \qquad (6.16)$$

on the variables r and $\dot{r}$ is then the same as for the one-dimensional motion if the substitutions $x \to r$, $\dot{x} \to \dot{r}$, and $V(x) \to V_{eff}(r)$ are made. Observe, however, that (6.16) refers to a mass point moving in two (or three) dimensions.

As in the one-dimensional case, we can now easily describe the motion of the particle by inspecting the graph of $V_{eff}(r)$. We choose, for example, the potential energy function $V(r)$ drawn in Figure 6.1a such that $V_{eff}(r)$ has a minimum at $r = r_0$. For the given value E of the total energy, the particle oscillates between the two values $r_1 = r_{min}$ and $r_2 = r_{max}$. These two radial distances define two circles about the force center in the plane of the orbit. Because of angular motion the particle's orbit lies inside the ring bounded by the two circles (see Figure 6.1b).

The orbit is not closed, in general. If the orbit is closed, the average frequency of revolution $\bar{f}_{rev}$ may be defined by

$$\bar{f}_{rev} = \frac{\text{number of oscillations necessary to close the orbit}}{\text{total time needed to close the orbit}},$$

and the ratio of the radial frequency of oscillation f_r to $\bar{f}_{rev}$, $f_r: \bar{f}_{rev}$, is a rational number. The value of this ratio depends on the force law and on the magnitudes of E and l.

There are only two force laws, namely, (1) the inverse square force law

$$f(r) = -\frac{k}{r^2} \quad \left(\text{or } V = -\frac{k}{r}\right), \quad \text{where} \quad E < 0,$$

and (2) the isotropic harmonic oscillator force law

$$f(r) = -kr \quad \left(\text{or } V = \tfrac{1}{2}kr^2\right)$$

for which the orbit is *always* closed independent of the values of E and l. The orbits are ellipses (or circles). But although in the first case the force center is located at a focal point of the ellipse (Kepler's first law), in the second case the force center is located at the center of the ellipse (see Figures 6.2a and 6.2b). In each of the two cases a physical quantity exists that is conserved in addition to and independently of E and l. The conserved quantity arises from the force law itself, and the associated symmetry is therefore called a dynamical symmetry.

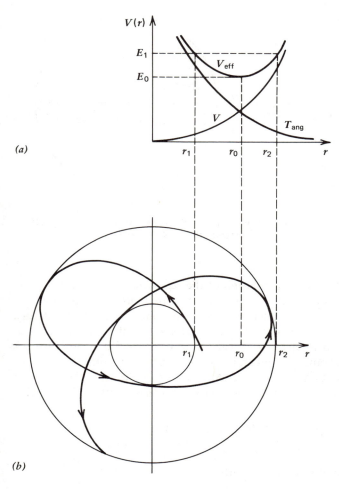

Figure 6.1 (a) Effective potential energy. (b) Particle orbit.

Dynamical symmetries play a crucial role in nuclear and elementary particle physics.

If the total energy takes the value $E = E_0 = V_{\text{eff}}(r_0)$, the orbit is a circle. Circular orbits are possible whenever V_{eff} has a minimum, and therefore they are not characteristic for the force law or the potential energy function $V(r)$.

We return to the theorem of energy conservation, Eq. 6.16. Performing another integration, we obtain the radial equivalent of (4.57),

$$t(r) = \int_{r_0}^{r} \frac{dr'}{\sqrt{\frac{2}{m}[E - V_{\text{eff}}(r')]}} = \int_{r_0}^{r} \frac{dr'}{\sqrt{\frac{2}{m}\left[E - V(r') - \frac{l^2}{2m}\frac{1}{r'^2}\right]}} \quad (6.17)$$

Assuming that $t(r)$ can be inverted, we arrive at $r = r(t)$.

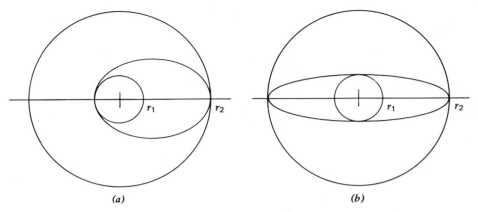

Figure 6.2 (a) Inverse square force law, $V(r) = -k/r \, (E < 0)$ closed orbit, $f_r : f_{rev} = 1 : 1$. (b) Isotropic harmonic oscillator, $V(r) = (1/2)kr^2$ closed orbit, $f_r : f_{rev} = 2 : 1$.

The energy integral is a formal solution of the equation of motion (6.8). The integral may be evaluated in closed form for only a limited number of force laws.[1] If $f(r)$ is of the form $f(r) = kr^n$, then for $n = 1$ (isotropic harmonic oscillator), $n = -2$ (gravitational and electrostatic forces), and $n = -3$ the integral leads to simple functions discussed in elementary calculus.[2] For $n = 5, 3, 0, -4, -5, -7$ and some fractional exponents, the integral leads to elliptic functions.

The energy equation may also be written in terms of the variables $u = 1/r$ and φ. If $\dot{r}$ is replaced by $-(l/m) \, du/d\varphi$ (see Eq. 6.11), we obtain from (6.16)

$$E = \frac{1}{2}\frac{l^2}{m}\left(\frac{du}{d\varphi}\right)^2 + V_{\text{eff}}\left(\frac{1}{u}\right)$$

Separation of variables and integration leads to the energy integral

$$\varphi(u) - \varphi_0 = \int_{u_0}^{u} \frac{du'}{\sqrt{\dfrac{2m}{l^2}\left(E - V_{\text{eff}}\left(\dfrac{1}{u'}\right)\right)}} = \int_{u_0}^{u} \frac{du'}{\sqrt{\dfrac{2m}{l^2}\left(E - V\left(\dfrac{1}{u'}\right)\right) - u'^2}}$$

$$(6.18)$$

which is the solution of the orbit equation (6.12). If $\varphi(u)$ can be inverted, we find the orbit $r = r(\varphi)$. The functions $r(t)$ or $r(\varphi)$ may be substituted into (6.7) to yield $\varphi = \varphi(t)$.

The general procedure for solving the equations of motion for central isotropic forces is summarized in Figure 6.3.

---

[1] For details see H. Goldstein, *Classical Mechanics,* 2nd ed., Addison-Wesley, Reading, Mass.: 1980.

[2] The corresponding quantum mechanical problems are also exactly solvable in terms of special mathematical functions.

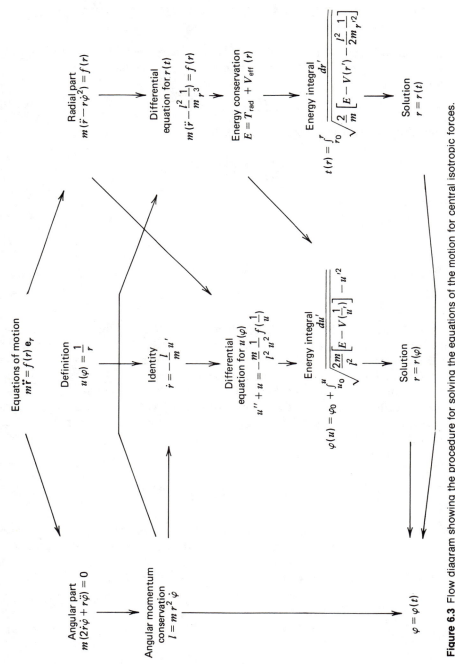

Figure 6.3 Flow diagram showing the procedure for solving the equations of the motion for central isotropic forces.

6.4 THE THREE-DIMENSIONAL ISOTROPIC HARMONIC OSCILLATOR

We proceed with the application of the formalism developed in the preceding two sections to special forces, and begin with the three-dimensional isotropic harmonic oscillator force

$$f(r) = -kr \tag{6.19}$$

The radial part of the equation of motion (6.8)

$$m\ddot{r} - \frac{l^2}{m}\frac{1}{r^3} = f(r) = -kr$$

may be brought into the form

$$r^3\ddot{r} + \omega_0^2 r^4 - \left(\frac{l}{m}\right)^2 = 0 \tag{6.20}$$

where

$$\omega_0^2 = \frac{k}{m} \tag{6.21}$$

The differential equation (6.20) is much more complicated than the three one-dimensional equations of the form (5.5) for the coordinates x, y, and z, and the statement in Section 6.1 that central forces should best be dealt with in plane polar coordinates seems to be contradicted. This exception to the rule is explained by the fact that for the isotropic harmonic oscillator force the equations of motion are intrinsically decoupled in Cartesian coordinates, whereas angular momentum conservation is directly exhibited in polar coordinates. For further details see Section 8.9.

The orbit equation (6.12)

$$u^3 u'' + u^4 - \left(\frac{m}{l\omega_0}\right)^2 = 0 \tag{6.22}$$

has the same mathematical structure as Eq. 6.20 because $V \propto r^2 \propto u^{-2}$ and $T_{\text{ang}} \propto r^{-2} \propto u^2$. We know from Section 5.5 that the orbit is an ellipse, and therefore, we are not interested in solving Eqs. 6.20 and 6.22. For the purpose of completeness, however, we shall state the solutions that satisfy the initial conditions $r(0) = r_{\text{max}}$ and $\varphi(0) = 0$:

$$r^2(t) = \frac{E}{k}\left[1 + \sqrt{1 - \left(\frac{l\omega_0}{E}\right)^2}\cos 2\omega_0 t\right], \tag{6.23}$$

$$r^2(\varphi) = \frac{l^2}{mE}\left[1 - \sqrt{1 - \left(\frac{l\omega_0}{E}\right)^2}\cos 2\varphi\right]^{-1} \tag{6.24}$$

Note that $\varphi \neq \omega_0 t$.

Because the orbit is closed for any choice of the initial conditions, the force law possesses a dynamical symmetry that guarantees the existence of another conserved dynamical quantity besides E and l. The total energy may be split up into the two parts

$$E_1 = \tfrac{1}{2}m\dot{x}^2 + \tfrac{1}{2}kx^2 \quad \text{and} \quad E_2 = \tfrac{1}{2}m\dot{y}^2 + \tfrac{1}{2}ky^2,$$

each being conserved separately. This is not the case for all the other isotropic central forces.

6.5 INVERSE SQUARE FORCE LAWS

A force that depends on the inverse square of the distance r between two particles is called inverse square force for short. Important examples are the static gravitational force between two mass points of masses m_{g1} and m_{g2}

$$f(r) = -G\frac{m_{g1}m_{g2}}{r^2} \tag{6.25}$$

and the electrostatic force between two point objects carrying electric charges q_1 and q_2

$$f(r) = k_e\frac{q_1q_2}{r^2} \tag{6.26}$$

The term *mass* has been introduced in two entirely different ways. In Eq. 4.2 the mass serves as a measure for the resistance of the change of the motion when a force acts on an object. In this case the mass is called inertial mass m_i. In Eq. 6.25 the mass is the source of the gravitational attraction, and m_g is called the gravitational mass. At this stage there is no logical reason to assert that m_g should be equal to m_i. However, experiments have shown [Eötvös (1889), Roll, Krotkov, and Dicke (1964), Braginski and Panov (1971), and recent lunar observations] that for all objects the masses m_g and m_i are equal to each other (by proper choice of the proportionality constant) to within an experimental error of less than $1 : 10^{12}$. This crucial result,

$$m_g = m_i \tag{6.27}$$

which is called the weak equivalence principle, is a cornerstone of general relativity, Einstein's theory of gravitation. The equivalence principle implies that all objects fall at the same rate in a gravitational force field. From now on we shall simply write m for both m_g and m_i.

The constant G appearing in (6.25) is called the gravitational constant; it was first measured by Cavendish. Even today its exact value is known to within a few digits only,

$$G = 6.6720(41) \times 10^{-11}\frac{Nm^2}{kg^2}$$

G expresses the strength of the gravitational interaction.

Although the gravitational force (6.25) is always attractive, the electrostatic force (6.26) may be attractive or repulsive depending on the sign of the electric charge q. The constant

$$k_e = \frac{1}{4\pi\varepsilon_0} = 8.9876 \times 10^9 \frac{Nm^2}{Coul^2}$$

expresses the electric coupling strength. For proton-proton interaction the electrostatic force is about 10^{36} times larger than the gravitational force.

We stated above that the force laws (6.25) and (6.26) hold for point particles. We shall now show that the same laws are valid for any two objects with a spherically symmetric mass (or charge) distribution.

We first calculate the force between a spherical shell of radius R and mass M with uniform surface mass density

$$\sigma = \frac{M}{4\pi R^2}$$

and a mass point of mass m placed outside the shell (see Figure 6.4). Because of symmetry the gravitational force df on m caused by a thin circular slice of the shell of mass dM is directed toward O, the center of the spherical shell. The magnitude of df is given by

$$df = G\frac{m\,dM}{\rho^2}\cos\varphi \qquad (6.28)$$

In order to express $\cos\varphi$ and dM as functions of ρ, we apply the law of cosines involving φ,

$$R^2 = \rho^2 - r^2 - 2\rho r \cos\varphi$$

which can be written

$$\cos\varphi = \frac{\rho^2 + r^2 - R^2}{2r\rho} \qquad (6.29)$$

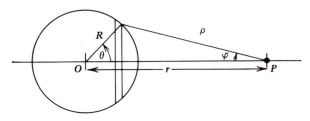

Figure 6.4 Diagram to aid the derivation of the force law between a mass point and a spherical shell.

From the law of cosines involving θ,

$$\rho^2 = R^2 + r^2 - 2Rr\cos\theta$$

we obtain after differentiation

$$\sin\theta\, d\theta = \frac{\rho\, d\rho}{rR} \tag{6.30}$$

The mass element dM is equal to

$$dM = \sigma\, dA = \sigma \cdot 2\pi R \sin\theta \cdot R\, d\theta$$

$$= 2\pi \frac{M}{4\pi R^2} \cdot R^2 \sin\theta\, d\theta$$

$$= \frac{1}{2} M \sin\theta\, d\theta$$

which because of (6.30) leads to

$$dM = \frac{M}{2rR} \rho\, d\rho \tag{6.31}$$

Substituting Eqs. 6.29 and 6.31 into Eq. 6.28, and integrating over ρ we obtain the result

$$f(r) = \int df = \int_{r-R}^{r+R} G \frac{m\left(\dfrac{M}{2rR}\rho\, d\rho\right)}{\rho^2} \cdot \frac{\rho^2 + r^2 - R^2}{2r\rho}$$

$$= G\frac{mM}{r^2}\frac{1}{4R}\int_{r-R}^{r+R}\left(1 + \frac{r^2 - R^2}{\rho^2}\right) d\rho$$

$$= G\frac{mM}{r^2} \tag{6.32}$$

because the last integral is equal to $4R$. We have recovered the inverse square force law, whereby the mass of the shell may be thought of being localized at the center O of the sphere. Because the force law (6.32) does not depend on the radius R of the shell, it must also hold for any number of concentric spherical shells, that is, for any body with spherically symmetric mass distribution around O. This result permits us to replace the sphere by a mass point (of the same mass M as that of the sphere) located at O, and to repeat the derivation of (6.32) with mass shells now centered around P. It follows that (6.32) holds for any two spherically symmetric mass distributions provided that they do not intersect each other.

Also, radial oscillation of the objects does not change the force law (6.32) (Birkhoff's theorem in general relativity). It is known that the best-observed oscillation of the sun (that with a period $T \approx 5$ min) proceeds radially with an amplitude of about 500 m. Even if this amplitude were not small compared with the radius of the sun ($R = 7 \times 10^8$ m), the oscillation would have no effect on the orbits of the planets.

The gravitational force between a uniform spherical mass shell and a mass point placed inside the shell is zero (see Problem 6.7a). It then follows that the gravitational force on a mass point of mass m inside a sphere of uniform mass density (radius R, mass M) is equal to

$$f(r) = -G\frac{Mm}{R^3}r \tag{6.33}$$

(see Problem 6.7b), which is an isotropic harmonic oscillator force.

6.6 KEPLER'S LAWS OF PLANETARY MOTION

One of the outstanding contributions to physics and astronomy was made by Tycho Brahe (1546–1601) who over a period of 20 years observed and recorded the orbits of the then known planets. The accuracy of his recorded data is remarkable insofar as he never used a telescope. Johannes Kepler (1571–1630) analyzed Brahe's data and arrived at the following laws:

1. The orbit of a planet is an ellipse with the sun at one focus.

2. The radius vector from the sun to the planet sweeps out equal areas in equal time intervals.

3. The square of the period of revolution divided by the cube of the semimajor axis has the same[3] value for all planets.

At an earlier stage Kepler conjectured that the planetary orbits are circles, but he soon had to reject this idea because of disagreements with Brahe's data, although the deviation from circular orbits is very small for most planets. It will be shown in the following section that the first and the third laws are direct consequences of the inverse square force law (6.32) discovered by Newton.

The second law is the theorem of angular momentum conservation in geometrical formulation. It is therefore valid for all central forces. In particular, the second law implies that the gravitational force is central. Referring to Figure 6.5, we have

$$\text{constant} = \frac{dA}{dt} = \frac{\frac{1}{2}r\,ds\sin\theta}{dt} = \frac{1}{2}rv\sin\theta$$

$$= \frac{1}{2m}m|\mathbf{r} \times \mathbf{v}| = \frac{1}{2m}l \tag{6.34}$$

The development of the explanation of planetary motion is an outstanding example of the development of new theories in physics: from observations via a geometrical description to the dynamical explanation. Other examples are given in Table 6.1.

[3]See, however, Section 7.6.

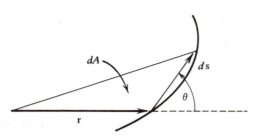

Figure 6.5 Diagram to aid the derivation of Kepler's second law.

Table 6.1 Some Theories of Physics in Developmental Overview

Subject Matter	Observation	Geometrical Description	Dynamical Explanation
Planetary motion	Orbits (Brahe)	Kepler's laws	Newton's law of gravitation
Electromagnetic field	Electric and magnetic phenomena (Thales of Miletus, Oersted,...)	Field lines (Faraday)	Maxwell's equations
Atomic structure	Rutherford's scattering experiments	Planetary model (Rutherford, Bohr)	Quantum mechanics
Spectral lines	Characteristic line spectra (Wollaston, Kirchhoff, Bunsen,...)	Rydberg-Ritz combination principle, application of group theory[a]	Quantum mechanics
Elementary particles	More than 200 known elementary particles	Unitary symmetries, quark model	Still unknown in spite of the progress made in recent years (unified field theories, quantum chromo dynamics)

[a] Historically, the application of group theory followed the discovery of quantum mechanics.

6.7 ORBITS IN THE INVERSE SQUARE FORCE FIELD

Let us write the inverse square force law in the form

$$f(r) = -k\frac{1}{r^2} \tag{6.35}$$

where

$$k = GMm \tag{6.36a}$$

for the gravitational force, and

$$k = -k_e q_1 q_2 \tag{6.36b}$$

for the electrostatic force. The force is attractive if $k > 0$, and repulsive if $k < 0$. The potential energy is given by

$$V(r) = -\frac{k}{r} \tag{6.37}$$

Inserting (6.35) into the differential equation (6.12) for the orbit we obtain

$$u'' + u = -\frac{m}{l^2}\frac{1}{u^2}f\left(\frac{1}{u}\right) = -\frac{m}{l^2}\frac{1}{u^2}(-ku^2) = \frac{mk}{l^2}$$

or

$$u'' + u = \frac{1}{\alpha}, \tag{6.38}$$

where we have set

$$\alpha = \frac{l^2}{mk} \tag{6.39}$$

The general solution of the inhomogeneous equation (6.38) is given by the sum of the general solution

$$u_{gh} = \beta \cos(\varphi + \varphi_0)$$

of the associated homogeneous equation, and a particular solution

$$u_{pi} = \frac{1}{\alpha}$$

of the inhomogeneous equation. Thus,

$$u(\varphi) = u_{gh} + u_{pi} = \beta \cos(\varphi + \varphi_0) + \frac{1}{\alpha},$$

and the equation for the orbit becomes

$$r(\varphi) = \frac{1}{u} = \frac{\alpha}{1 + \alpha\beta \cos(\varphi + \varphi_0)}$$

Table 6.2 Conic Sections

	Circle	Ellipse	Parabola	Hyperbola	Straight Line
Cartesian coordinates Origin O	$x^2 + y^2 = a^2$	$\dfrac{x^2}{a^2} + \dfrac{y^2}{G^2} = 1$	$y^2 = \overset{+}{_{(-)}} 4ax$	$\dfrac{x^2}{a^2} - \dfrac{y^2}{b^2} = 1$	$y = 0$
Distance $OF = OF'$	0	$c = \sqrt{a^2 - b^2}$	a	$c = \sqrt{a^2 + b^2}$	0
Eccentricity ε	0	$\dfrac{1}{a}\sqrt{a^2 - b^2} < 1$	1	$\dfrac{1}{a}\sqrt{a^2 + b^2} > 1$	∞
Polar coordinates Origin F'	$r = a$	$r = \dfrac{a(1 - \varepsilon^2)}{1 - \varepsilon \cos\varphi}$	$r = \dfrac{2a}{1 - \cos\varphi}$	$r = \dfrac{a(\varepsilon^2 - 1)}{1 - \varepsilon \cos\varphi}$	$\varphi = 0$
			(right branch)	$\left(\text{right branch, } \cos\varphi < \dfrac{1}{\varepsilon}\right)$ (right branch)	
Particles in an inverse square force field: Energy	$E = E_{\min} = -\dfrac{mk^2}{2l^2} < 0$	$E_{\min} < E < 0$	$E = 0$	$0 < E$	$E = \infty \quad \text{or} \quad l = 0$
Attractive force: force center at	0	F' or F	F'	F'	0
Repulsive force: force center at	—	—	F	F	0

Without loss of generality we choose the coordinate axes such that $\varphi_0 = \pi$. Then we finally obtain

$$r(\varphi) = \frac{\alpha}{1 - \alpha\beta\cos\varphi} \qquad (6.40)$$

which is the equation of a conic section in polar coordinates (see Table 6.2). Because the origin of the coordinate system is located at a focus, the force center will be at the same focus of the conic section; this proves Kepler's first law.

Before discussing the result (6.40) we take a brief look at the effective potential energy

$$V_{\text{eff}}(r) = T_{\text{ang}} + V(r) = \frac{l^2}{2m}\frac{1}{r^2} - \frac{k}{r} \qquad (6.41)$$

which is drawn in Figures 6.6a and 6.6b for the two cases $k > 0$ and $k < 0$. For attractive forces, $k > 0$, the orbit may be any of the five conic sections depending on the value of the total energy. The orbit is:

(a)   a circle if $E = E_0 = E_{\text{min}} < 0$,
(b)   an ellipse if $E_0 < E < 0$ (planetary motion),
(c)   a parabola if $E = 0$,
(d)   a hyperbola if $E > 0$ (some comets),
(e)   a straight line if $l = 0$ or $E = \infty$ (see Eq. 6.51 below).

Circular, parabolic, and straight line orbits are included here for mathematical completeness rather than for their practical realizability. For repulsive forces only hyperbolic orbits are possible because the total energy is always positive (e.g., α particle scattering).

We return to Eq. 6.40. Comparing the coefficients of (6.40) with those of the general formula of a conic section (see Table 6.2) we obtain

$$\alpha = a(1 - \varepsilon^2) \qquad (6.42)$$

and

$$\alpha\beta = \varepsilon \qquad (6.43)$$

More information is needed to relate the constants a, ε, and β to the dynamical quantities E and l. This may be accomplished, for example, by evaluating the energy integral (6.18) for the potential energy function (6.37). Here we pursue another possibility, namely, to employ the theorem of energy conservation directly.

A point on an orbit for which the distance between that point and the force center is an extremum, is called an apsis. There are only two apsides for elliptical orbits. If the sun is the force center, the point P closest to the sun is called

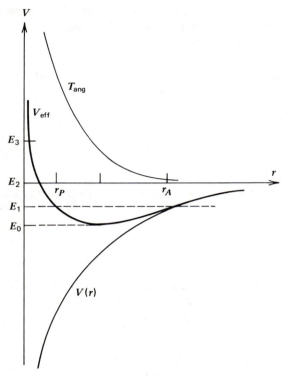

(a)

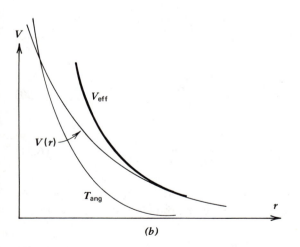

(b)

Figure 6.6 (a) Effective potential energy for the attractive inverse square force field, $k > 0$. (b) Effective potential energy for the repulsive inverse square force field, $k < 0$.

perihelion, the point A farthest away from the sun is called aphelion.[4] For orbits around the earth the corresponding points are called perigee and apogee.

The coordinate r_P of P is found by setting $\varphi = \pi$:

$$r_P = a(1 - \varepsilon) = \frac{\alpha}{1 + \alpha\beta} \tag{6.44}$$

Similarly, for $\varphi = 0$ one obtains

$$r_A = a(1 + \varepsilon) = \frac{\alpha}{1 - \alpha\beta} \tag{6.45}$$

Because the radial velocity vanishes at both P and A, $\dot{r} = 0$, the total energy is equal to

$$E = \frac{l^2}{2m}\frac{1}{r_P^2} - \frac{k}{r_P} = \frac{l^2}{2m}\frac{1}{r_A^2} - \frac{k}{r_A} \tag{6.46}$$

Equations 6.39 and 6.42 through 6.46 may be combined to yield other relations between the various quantities. We state some of these equations and leave the proof to the reader, because only elementary algebra is needed.

$$\left.\begin{matrix} r_A \\ r_P \end{matrix}\right\} = -\frac{k}{2E}\left\{1 \pm \sqrt{1 + \frac{2l^2 E}{mk^2}}\right\} \tag{6.47}$$

$$a = \frac{1}{2}(r_A + r_P) = -\frac{k}{2E} \tag{6.48}$$

$$\varepsilon = \sqrt{1 - \frac{\alpha}{a}} = \sqrt{1 + \frac{2l^2 E}{mk^2}} \tag{6.49}$$

$$\beta = \frac{\varepsilon}{\alpha} = \frac{mk}{l^2}\sqrt{1 + \frac{2l^2 E}{mk^2}} \tag{6.50}$$

$$b = a\sqrt{1 - \varepsilon^2} = \sqrt{r_A r_P} = \sqrt{\frac{l^2 a}{mk}} = \sqrt{-\frac{l^2}{2mE}} \tag{6.51}$$

Equation 6.49 proves quantitatively that, for given l, the value of E determines the type of the conic section. For circular orbits, $\varepsilon = 0$, E is given by

$$E = -\frac{mk^2}{2l^2} \tag{6.52}$$

[4] Peri (Greek): near; apo (Greek): away from; helios (Greek): sun; ge (Greek): earth.

Although the length a of the semimajor axis depends solely on E, the magnitude b of the semiminor axis depends on both E and l.

Kepler's third law follows now directly from Eq. 6.34. If we write the (constant) areal velocity dA/dt in terms of the area $A = \pi ab$ of the ellipse and the period T, we obtain

$$\frac{l}{2m} = \frac{dA}{dt} = \frac{A}{T} = \frac{\pi ab}{T}$$

which, after elimination of b using (6.51) becomes

$$\frac{l}{2m} = \frac{\pi a}{T} \sqrt{\frac{l^2 a}{mk}}$$

Squaring the last equation and rearranging the terms we obtain

$$\frac{T^2}{a^3} = (2\pi)^2 \frac{m}{k} = \frac{(2\pi)^2}{GM} \tag{6.53}$$

This ratio has the same value for all planets because it does not contain the mass of the planet.[5] Equation 6.53 can be used to determine the solar mass from the orbital data of the planets and the value of the gravitational constant.

Like the isotropic harmonic oscillator force (6.19), the inverse square force possesses a dynamical symmetry. The Laplace-Runge-Lenz vector $\mathbf{Z}$ defined by

$$\mathbf{Z} = \frac{\mathbf{r}}{r} - \frac{1}{km}\mathbf{p} \times \mathbf{l} \tag{6.54}$$

is a constant of the motion (see Problem 6.10a):

$$\dot{\mathbf{Z}} = 0 \tag{6.55}$$

$\mathbf{Z}$ may be written in the form (see Problem 6.10b)

$$\mathbf{Z} = \varepsilon \frac{\mathbf{r}_A}{r_A} \tag{6.56}$$

where $\mathbf{r}_A$ is the vector from the force center to the apsis A. The Laplace-Runge-Lenz vector vanishes for circular orbits. Although $\mathbf{Z}$ has three (constant) components, only one of them is independent of E and l.

We conclude this section with a brief comment on the central force problem in quantum mechanics. For central isotropic forces the eigenvalues of the energy operator depend on the two quantum numbers n and l, whereas for the isotropic harmonic oscillator and the inverse square force laws (both possessing dynamical symmetries) the energy spectrum is described by only one quantum number, namely, n. Dynamical symmetries give rise to degeneracies of the eigenvalue spectrum in quantum mechanics.

[5]See, however, Section 7.6.

6.8 SCATTERING IN A CENTRAL FORCE FIELD

6.8.1 The Scattering Cross Section

One of the main problems of physics is the determination of the exact form of the force between two particles or systems of particles from observational data. A knowledge of the force may provide information about the composition and structure of the physical system.

For interactions between two macroscopic objects the force law can be found from the equations of motion if the orbits of the objects are known simply by differentiating the functions $r(t)$ (Eq. 4.4), or $r(\varphi)$ (Eq. 6.12). This method fails, however, for both theoretical and experimental reasons when microphysical particles are involved. First, the framework of classical mechanics must be replaced by quantum mechanics where the concept of orbit loses its meaning. Second, there is no experimental procedure by which a microscopic particle can be observed over very small distances (e.g., an α particle while passing the nucleus of an atom). Third, even if observational techniques were refined sufficiently to determine the particle's position, a change of its momentum and hence its orbit would be unavoidable during the process of measurement because of the interaction (collision) of the observing light ray (photon) with the particle.

What can be measured with sufficient accuracy are the momenta of the particle before and after the interaction has taken place. Particles that are aimed toward a force center and deflected by it are said to be scattered. Because the orbits of individual particles cannot be traced, statistical methods must be applied to describe the overall behavior of the scattered particles. One is then faced with the problem of drawing conclusions about the interaction from the statistical distribution of the change of the motion of the particles.

The concept of cross section has turned out to be very fruitful. It is a measure for the probability, in units of area, that a certain reaction takes place, for example, scattering into a solid angle, absorption of a particle, creation of new particles, and so on. An understanding of cross sections in classical mechanics is of paramount importance before dealing with equivalent problems in quantum physics.

The solid angle Ω is defined as the ratio of the area A of a portion of a sphere of radius r to the square of the radius:

$$\Omega = \frac{A}{r^2}$$

In spherical polar coordinates the infinitesimal solid angle $d\Omega$ is given by

$$d\Omega = d\varphi \sin \theta \, d\theta$$

such that for a full sphere, as expected,

$$\frac{4\pi r^2}{r^2} = \Omega = \int_0^{2\pi} d\varphi \int_0^{\pi} \sin \theta \, d\theta = 2\pi \times 2 = 4\pi$$

The particle current density I is the number N of particles traversing an area A perpendicular to the momentum during the time interval Δt:

$$I = \frac{N}{A \, \Delta t} \tag{6.57}$$

If N_0 denotes the number of incident particles, the corresponding particle current density is

$$I_0 = \frac{N_0}{A \, \Delta t} \tag{6.58}$$

In order to describe the position of the scattered particle, we choose spherical polar coordinates with the origin attached to the target particle and with the z-axis aligned in the direction of the incoming free particle. Let $N(\theta, \varphi) \, d\Omega$ be the number of particles scattered into the direction $\{\theta, \varphi\}$ within the solid angle $d\Omega$, and let K be the number of target particles. Then

$$n(\theta, \varphi) \, d\Omega = \frac{N(\theta, \varphi) \, d\Omega}{K \, \Delta t} \tag{6.59}$$

is the number of particles scattered within $d\Omega$ per target particle and unit time. The number of particles per unit area is

$$\eta = \frac{K}{A} \tag{6.60}$$

The scattering differential cross section $\sigma(\theta, \varphi)$ is defined by

$$\sigma(\theta, \varphi) \, d\Omega = \frac{n(\theta, \varphi) \, d\Omega}{I_0}, \tag{6.61}$$

which because of Eqs. 6.58, 6.59, and 6.60 is equal to

$$\sigma(\theta, \varphi) \, d\Omega = \frac{n(\theta, \varphi) \, d\Omega}{N_0 \eta} \tag{6.62}$$

The cross section $\sigma(\theta, \varphi)$ represents an area through which the particles must pass in order to be scattered into the spherical angle $d\Omega = d\varphi \sin \theta \, d\theta$.

In (6.62), $\sigma(\theta, \varphi)$ is expressed in terms of quantities that can be *measured* directly. We now wish to find an equation that relates $\sigma(\theta, \varphi)$ to convenient geometrical and/or dynamical quantities describing the interaction. We restrict ourselves to the following case:

(a) The target particle (scattering center) is assumed to be fixed in space. This may be realized if the mass of the target particle is much larger than the mass of the incident particle.

(b) There is only single scattering, or, in other words, multiple (repeated) scattering is assumed to be improbable. The target particles may be arranged in a thin layer (foil).

(c) The force of interaction is central isotropic. Then the scattering cross section is independent of the azimuth angle φ. Integrating Eq. 6.62 over φ we obtain

$$2\pi\sigma(\theta)\sin\theta\,d\theta = \frac{2\pi N(\theta)\sin\theta\,d\theta}{N_0\eta} \tag{6.63}$$

A convenient quantity used to describe collisions is the impact parameter s. It is geometrically defined as the shortest distance between the target particle and the incident particle that they would attain *if the interaction were absent*. The impact parameter can also be expressed in terms of the dynamical quantities linear momentum $\mathbf{p}$ of the incident *free* particle and the (conserved) angular momentum $\mathbf{l}$ of the particle measured from the force center:

$$s = \frac{|\mathbf{l}|}{|\mathbf{p}|} \tag{6.64}$$

(see Problem 6.12). Particles that pass through the ring between the two circles of radii s and $s + ds$, respectively, are scattered into the direction θ within $d\theta$ (see Figure 6.7). Therefore, the differential cross section is simply equal to

$$2\pi\sigma(\theta)\sin\theta\,d\theta = 2\pi s\,ds \tag{6.65}$$

It is not possible to deduce the force law from the measured cross section (6.63). The best that one can do is to guess a force law, then calculate the cross section from Eq. 6.65, and see whether the calculated function $\sigma(\theta)$ agrees with the experimental data. If the agreement is good, one has a reason to believe that the force law is the "correct" one.

The total scattering cross section

$$\sigma_{\text{total}} = \int\sigma(\theta,\varphi)\,d\Omega \tag{6.66}$$

is the area through which a particle must pass to be scattered in any direction.

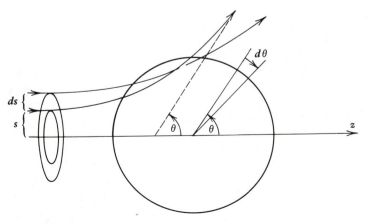

Figure 6.7 Impact parameter and scattering angle.

Table 6.3 Range of Forces Versus Total Cross Section

Short-range force:	Force which decreases faster than r^{-2} as $r \to \infty$.	σ_{total} = finite
Long-range force:	Force which is not a short range force.	σ_{total} = infinite

According to whether σ_{total} is finite or infinite, the force of interaction is called of finite or infinite range (see Table 6.3).

6.8.2 Scattering of Hard Spheres

We shall now discuss two examples. We begin with the mathematically simple case of hard sphere scattering.

Let a be the radius of the target particle and b that of the incident particle. Inspecting Figure 6.8, we find the identities $\alpha + \varphi = \pi$ and $2\alpha + \theta = \pi$, hence

$$\varphi = \frac{\theta}{2} + \frac{\pi}{2}$$

The impact parameter is then given by

$$s = (a + b)\sin \varphi = (a + b)\sin\left(\frac{\theta}{2} + \frac{\pi}{2}\right)$$

$$= (a + b)\cos \frac{\theta}{2}, \tag{6.67}$$

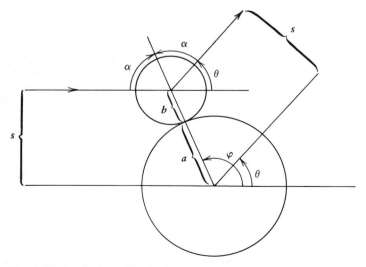

Figure 6.8 Scattering of hard spheres.

and its differential is equal to

$$ds = -(a+b)\frac{1}{2}\sin\frac{\theta}{2}d\theta \qquad (6.68)$$

Substituting Eqs. 6.67 and 6.68 into Eq. 6.65 we arrive at

$$2\pi\sigma(\theta)\sin\theta\,d\theta = 2\pi s\,ds$$

$$= 2\pi(a+b)\cos\frac{\theta}{2}\cdot(-1)(a+b)\frac{1}{2}\sin\frac{\theta}{2}d\theta$$

$$= -\frac{\pi}{2}(a+b)^2\sin\theta\,d\theta \qquad (6.69)$$

Because the differential cross section

$$|\sigma(\theta)| = \tfrac{1}{4}(a+b)^2$$

is a constant, the scattering of hard spheres is isotropic. The minus sign in (6.69) is of little importance. It arises from the fact that θ increases as s decreases.

The total scattering cross section (6.66) becomes

$$|\sigma_{\text{total}}| = \left|\int_0^\pi 2\pi\sigma(\theta)\sin\theta\,d\theta\right|$$

$$= \frac{\pi}{2}(a+b)^2\int_0^\pi \sin\theta\,d\theta = \pi(a+b)^2 \qquad (6.70)$$

As expected, σ_{total} is finite because the force is of short range; σ_{total} is equal to the area of a circle of radius $(a+b)$ (see Figure 6.9). The incident particle is always scattered if $s < (a+b)$.

6.8.3 Scattering in an Inverse Square Force Field
The classical example for particle scattering in an inverse square force field is Rutherford's experiment (performed with Geiger and Marsden). In order to test

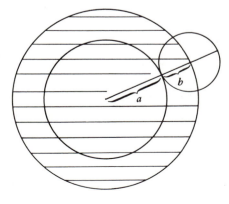

Figure 6.9 Total cross section for hard sphere scattering.

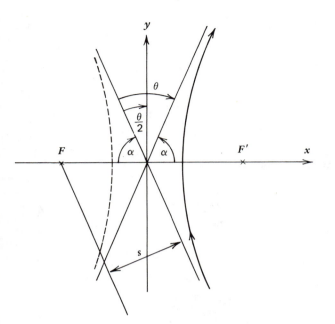

Figure 6.10 Scattering in an inverse square force field.

Thomson's model of the atom (plum pudding model), Rutherford aimed α particles at a thin gold foil and measured the number of scattered α particles as a function of the deflection angle. The outcome of this experiment, in particular the observed deflection of α particles by more than 90°, disagreed with Thomson's model. Rutherford, therefore, proposed another model of the atom, the planetary model. He conjectured that negatively charged electrons move about the positively charged massive nucleus like planets about the sun. In this model[6] deflections of α particles aimed at an atom arise mainly from the repulsive Coulomb interaction of the α particle with the nucleus, whereas electrons have only little effect on the motion of the α particle because of the small mass ratio $m_{el} : m_{\alpha}$.

The electric charges of the α particle and the nucleus of an atom with atomic number Z are $q_1 = +2e$ and $q_2 = +Ze$, respectively, where $e = 1.6 \times 10^{-19}$ Coulomb is the elementary charge; $Z = 79$ for gold. The constant k, Eq. 6.36b, appearing in the electrostatic force law (6.35), is then equal to

$$k = -k_e 2Ze^2 \tag{6.71}$$

Because the total energy of the α particle is positive, the orbit of the α particle is a hyperbola. The slope of the asymptotes is given by

$$\tan \alpha = \pm \frac{b}{a}$$

[6]Rutherford's atom is unstable according to the classical theory of electricity. Because the electrons are continuously accelerated (centripetal acceleration), they should continuously radiate energy, forcing the atom to collapse. Quantum mechanics explains why this does not happen.

Inspecting Figure 6.10 (see also Table 6.2) we deduce

$$\cot \frac{\theta}{2} = \cot(90° - \alpha) = \tan \alpha = \frac{b}{a}$$

$$= \sqrt{\frac{b^2 + a^2}{a^2} - 1} = \sqrt{\varepsilon^2 - 1}$$

which because of (6.49) can be written

$$\cot \frac{\theta}{2} = \sqrt{\frac{2El^2}{mk^2}} \qquad (6.72)$$

Eliminating l using (6.64), we obtain

$$\cot \frac{\theta}{2} = \sqrt{\frac{2E(mv_0 s)^2}{mk^2}} = 2\frac{E}{k}s,$$

hence,

$$s = \frac{k}{2E} \cot \frac{\theta}{2}$$

and

$$ds = -\frac{k}{2E} \frac{1}{\sin^2 \frac{\theta}{2}} d\left(\frac{\theta}{2}\right) \qquad (6.73)$$

The differential cross section (6.65) is then equal to

$$2\pi\sigma(\theta)\sin\theta \, d\theta = 2\pi \frac{k}{2E} \cot \frac{\theta}{2} \times \left(-\frac{k}{2E}\right) \frac{1}{\sin^2 \frac{\theta}{2}} d\left(\frac{\theta}{2}\right)$$

$$= -2\pi\left(\frac{k}{4E}\right)^2 \frac{2\sin \frac{\theta}{2} \cos \frac{\theta}{2}}{\sin^4 \frac{\theta}{2}} d\theta$$

$$= -2\pi\left(\frac{k}{4E}\right)^2 \frac{1}{\sin^4 \frac{\theta}{2}} \sin\theta \, d\theta \qquad (6.74)$$

from which follows Rutherford's scattering formula

$$|\sigma(\theta)| = \left(\frac{k}{4E}\right)^2 \frac{1}{\sin^4 \dfrac{\theta}{2}} \tag{6.75}$$

The following conclusions can be drawn from Eq. 6.75:

(a) $\sigma(\theta)$ is independent of the sign of k. Attractive and repulsive forces yield the same differential cross section.

(b) $\sigma(0) = \infty$. Particles that are not scattered at all must pass the nucleus at an infinitely large[7] impact parameter s. Because of the strong singularity at $\theta = 0$, we also have $\sigma_{\text{total}} = \infty$. Therefore, Coulomb forces are of infinite range.

(c) As the scattering angle increases, $\sigma(\theta)$ decreases monotonically to its minimum value at $\theta = 180°$. Because

$$\sigma(180°) = \left(\frac{k}{4E}\right)^2 \neq 0$$

backward scattering occurs. The distance r_{min} of closest approach is obtained from the energy equation

$$E = \frac{1}{2} m v_0^2 = \frac{k}{r_{\text{min}}}$$

We find

$$r_{\text{min}} = \frac{k}{\frac{1}{2} m v_0^2} = \frac{k}{E}$$

and

$$\sigma(180°) = \left(\frac{r_{\text{min}}}{4}\right)^2$$

(d) $\sigma(\theta)$ depends on only one dynamical variable, namely, E. For other interactions the differential cross section may depend on the angular momentum also. Investigations of the dependence of σ on E and l have led to valuable insights into the nature of the interactions between elementary particles.

It is interesting to note that the quantum mechanical calculation of the cross section leads to the same result (6.75) for low energies of the α particle. The fundamental constant of quantum mechanics, Planck's constant $h = 6.626 \times 10^{-34}$ Js, does not appear, which is another special feature of the inverse square force law.

---

[7]Here, the nucleus is considered without its surrounding electrons which produce a shielding effect. See also Section 7.7.

PROBLEMS

6.1 Derive the theorem of energy conservation (6.14) directly from the equations of motion (6.4) and (6.5) in polar coordinates without reference to the conservation of angular momentum.

6.2 Assume that the orbit function $r(\varphi)$ has a maximum (or minimum) at $\varphi = 0 : r_m = r(0)$. Show that $r(\varphi) = r(-\varphi)$, that is, the orbit possesses reflection symmetry.

6.3 (a) Find the force law if the orbit of a particle is given by

$$r = A(\cos n\varphi)^{-1/n}, \quad n \neq 0, \quad n \neq 1$$

(b) A particle moves on a circle that passes through the force center. The diameter of the circle is A. Find the force law.

6.4 (a) Find the force law if the orbit of a particle is given by

$$r = A(1 + \cos n\varphi)^{-1/n}, \quad n \neq 0, \quad n \neq 2$$

(b) A particle moves on a cardioid $r(\varphi) = A(1 + \cos\varphi)$. Find the force law.

6.5 Apply Eq. 6.12 to derive the force law if the orbit is a circle about the force center. What is strange about this result? This problem shows that (6.12) is meaningful only if u is a function of φ.

6.6 Solve the equation of motion for the inverse cube force law $f(r) = -kr^{-3}$. Consider the three cases k larger than, equal to, and smaller than l^2/m.

6.7 (a) Prove that the gravitational force between a uniform spherical mass shell and a mass point placed arbitrarily inside the shell is zero.
(b) Derive Eq. 6.33.

6.8 Derive Eqs. 6.47 through 6.51.

6.9 The maximum and minimum velocities of a planet moving about a star are $v_P = v_0 + v_1$ and $v_A = v_0 - v_1$.
(a) Find the eccentricity of the orbit in terms of v_0 and v_1.
(b) Show that $v_P v_A = (2\pi a/T)^2$ where T is the period of revolution and a is the magnitude of the semimajor axis.

6.10 (a) Show that the Laplace-Runge-Lenz vector $\mathbf{Z}$ is a constant of the motion.
(b) Derive Eq. 6.56.

6.11 For the earth of mass M and radius R, show that the following time intervals are the same:
(a) The period of oscillation of a pendulum of length R at the surface of the earth for small amplitudes (see also Problem 5.7). (See Figure 6.11a.)
(b) The period of oscillation of a mass point moving through a straight hole drilled through the earth, about the equilibrium point. (See Figure 6.11b.)

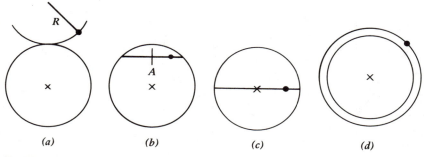

(a)   (b)   (c)   (d)

Figure 6.11 Problem 6.11.

(c) The period of oscillation of a mass point moving through a straight hole drilled through the center of the earth. (See Figure 6.11c.)

(d) The period of revolution of a satellite moving around the earth on a circular orbit of radius R. (See Figure 6.11d.)

6.12 Derive Eq. 6.64.

6.13 Derive Eq. 6.72 from the energy integral (6.18). Replace φ by θ.

6.14 Particles are scattered in an inverse cube force field

$$f(r) = kr^{-3}, \quad k > 0.$$

(a) Show that the impact parameter is given by

$$s = \sqrt{\frac{k}{2E}} \cdot \frac{1 - x}{\sqrt{(1 - x)^2 - 1}} \qquad \text{where} \qquad x = \frac{\theta}{\pi}$$

(b) Derive the differential cross section.

(c) Derive a formula for $\sigma(\theta)$ for small angles θ. See Problem 6.6.

6.15 (a) A particle of energy E moves from a region I where $V(\mathbf{r}) = 0$ to a region II where $V(\mathbf{r}) = -V_0 = \text{constant}$ (see Figure 6.12). Show that Snell's law of refraction holds in the form

$$\frac{\sin \theta}{\sin \theta'} = n = \sqrt{\frac{E + V_0}{E}}$$

(b) Particles are scattered by a spherically symmetric potential well

$$V(r) = \begin{cases} 0 & \text{if} \quad r \geqslant a \\ -V_0 & \text{if} \quad r < a \end{cases}$$

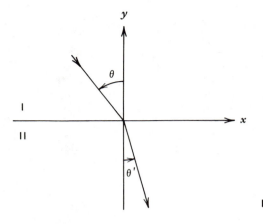

Figure 6.12 Problem 6.14.

Show that the impact parameter is given by

$$s(\theta) = a\sqrt{\frac{n}{2}} \cdot \frac{\sin\dfrac{\theta}{2}}{\sqrt{\dfrac{n^2+1}{2n} - \cos\dfrac{\theta}{2}}}$$

(c) Find the differential scattering cross section. Is backward scattering possible? Introduce the constant $b = (n^2 + 1)/2n$.

(d) Find the total scattering cross section. σ_{total}. Investigate the special cases $n = 1$ ($V_0 = 0$ or $E = \infty$) and $n = \infty$ ($V_0 = \infty$ or $E = 0$).

chapter

7 Many Particle Systems

Patience, and shuffle the cards.

CERVANTES

7.1 INTRODUCTION

In this chapter we shall investigate the general behavior of systems of N particles.

The ith particle is described by the mass m_i, the position vector $\mathbf{r}_i$ and its time derivatives velocity $\mathbf{v}_i = \dot{\mathbf{r}}_i$, and acceleration $\mathbf{a}_i = \ddot{\mathbf{r}}_i$. The force $\mathbf{F}_i$ acting on the ith mass point is composed of the external force $\mathbf{F}_i^{(e)}$ and the internal force

$$\mathbf{F}_i^{(i)} = \sum_{\substack{j=1 \\ j \neq i}}^{N} \mathbf{F}_{ji} \tag{7.1}$$

which is the resultant of all forces $\mathbf{F}_{ji}$ of the jth particle ($j = 1, 2, \ldots N, \ j \neq i$) on the ith particle. Thus,

$$\mathbf{F}_i = \mathbf{F}_i^{(e)} + \mathbf{F}_i^{(i)} = \mathbf{F}_i^{(e)} + \sum_{\substack{j=1 \\ j \neq i}}^{N} \mathbf{F}_{ji} \tag{7.2}$$

The equations of motion

$$m_i \ddot{\mathbf{r}}_i = \mathbf{F}_i = \mathbf{F}_i^{(e)} + \sum_{\substack{j=1 \\ j \neq i}}^{N} \mathbf{F}_{ji}, \quad i = 1, 2, \ldots N, \tag{7.3}$$

represent a system of N coupled vector differential equations because the internal force $\mathbf{F}_{ji} = -\mathbf{F}_{ij}$ affects both the ith and the jth particle. Because of the complexity of the equations, we do not expect that exact solutions $\mathbf{r}_i(t)$ of (7.3) can be found for large values of N. In Section 7.6 we shall outline the general solution procedure for the two-particle system. For the three-particle problem exact solutions exist only if very strong conditions are imposed on the masses and on the initial values of the coordinates and velocities. However, progress on the three body problem has been made in recent years. As N increases, Eqs. 7.3 become complicated, and approximation methods (perturbation theory) must be applied. For systems such as gases or galaxies, which consist of an extremely large number of particles, the "microscopic" coordinate vectors $\mathbf{r}_i$ are no longer meaningful because, for example, the orbit $\mathbf{r}_i(t)$ of a molecule cannot be observed. Therefore, one introduces a small number of "macroscopic" observable quantities that describe the overall behavior of the system. Some of these macroscopic quantities will be introduced and explored in the following sections.

7.2 CENTER OF MASS

The center of mass vector $\mathbf{R}$ is defined as the average position vector of all mass points weighted by the mass:

$$\mathbf{R} = \frac{1}{M} \sum_{i=1}^{N} m_i \mathbf{r}_i \tag{7.4}$$

where

$$M = \sum_{i=1}^{N} m_i \tag{7.5}$$

is the total mass of the system. The velocity and acceleration vectors of the center of mass are given by

$$\mathbf{v}_{CM} = \dot{\mathbf{R}} = \frac{1}{M} \sum_{i=1}^{N} m_i \dot{\mathbf{r}}_i = \frac{1}{M} \sum_{i=1}^{N} m_i \mathbf{v}_i \tag{7.6}$$

and

$$\mathbf{a}_{CM} = \ddot{\mathbf{R}} = \frac{1}{M} \sum_{i=1}^{N} m_i \ddot{\mathbf{r}}_i = \frac{1}{M} \sum_{i=1}^{N} m_i \mathbf{a}_i, \tag{7.7}$$

respectively. The last two equations are often used in the form

$$M\dot{\mathbf{R}} = \sum_{i=1}^{N} m_i \dot{\mathbf{r}}_i \tag{7.8}$$

and

$$M\ddot{\mathbf{R}} = \sum_{i=1}^{N} m_i \ddot{\mathbf{r}}_i \tag{7.9}$$

Because the right side of (7.8) is the total momentum of the system,

$$\mathbf{P} = \sum_{i=1}^{N} \mathbf{p}_i = \sum_{i=1}^{N} m_i \mathbf{v}_i, \tag{7.10}$$

it follows that

$$\mathbf{P} = M\dot{\mathbf{R}} \tag{7.11}$$

The total momentum of the system is equal to the total mass multiplied by the center of mass velocity.

7.3 FORCE AND MOMENTUM

Although Newton's third law, by its nature, is not needed in single particle dynamics, it is of central importance in the dynamics of many particle systems. We restate the third law in its two versions:

(a)   weak form: $\mathbf{F}_{ij} = -\mathbf{F}_{ji}$ (7.12)

(b)   strong form: $\mathbf{F}_{ij} = -\mathbf{F}_{ji}$ parallel to $\mathbf{r}_i - \mathbf{r}_j$ (7.13)

We shall see that the difference between the two forms is of minor consequence for rectilinear motion, whereas it becomes significant for rotational motion.

The total force acting on the system is defined by

$$\mathbf{F} = \sum_{i=1}^{N} \mathbf{F}_i = \sum_{i=1}^{N} \left(\mathbf{F}_i^{(e)} + \sum_{\substack{j=1 \\ j \neq i}}^{N} \mathbf{F}_{ji} \right) \tag{7.14}$$

Because of (7.12), the double sum $\Sigma_i \Sigma_j F_{ij}$ extended over all pairs of internal forces is identically zero, and the total force reduces to a sum over the external forces only:

$$\mathbf{F} = \sum_{i=1}^{N} F_i^{(e)} \tag{7.15}$$

Combining Eqs. 7.3, 7.9, 7.11, and 7.15, we obtain

$$\dot{\mathbf{P}} = M\ddot{\mathbf{R}} = \sum_{i=1}^{N} m_i \ddot{\mathbf{r}}_i = \sum_{i=1}^{N} \mathbf{F}_i = \mathbf{F} \tag{7.16}$$

Internal forces do not affect the motion of the center of mass. If the total force vanishes, $\mathbf{F} = 0$, which is, for example, the case if $\mathbf{F}_i = 0$ for all i, it follows from (7.8), (7.11), and (7.16) that

$$M\dot{\mathbf{R}} = \mathbf{P} = \text{constant} \tag{7.17}$$

This important statement constitutes the theorem of momentum conservation of a system. In this case the center of mass moves uniformly on a straight line, whereas the individual particles may follow complicated orbits caused by the mutual interaction between the particles. We see now from (7.16) and (7.17) that the center of mass is a meaningful macroscopic quantity.

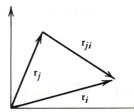

Figure 7.1 The relative position vector $\mathbf{r}_{ji}$.

If the internal forces $\mathbf{F}_{ji}$ are conservative, potential energy functions $V_{ji} = V_{ij}$ exist[1] such that

$$\mathbf{F}_{ji} = -\nabla_i V_{ji} \quad \text{and} \quad \mathbf{F}_{ij} = -\nabla_j V_{ij}, \tag{7.18}$$

where

$$\nabla_i = \mathbf{i}\frac{\partial}{\partial x_i} + \mathbf{j}\frac{\partial}{\partial y_i} + \mathbf{k}\frac{\partial}{\partial z_i} \tag{7.19}$$

is the del operator associated with the coordinates of the ith particle. Because of (7.12) and (7.18), it follows that V_{ij} is a function of the difference

$$\mathbf{r}_{ji} = \mathbf{r}_i - \mathbf{r}_j = -\mathbf{r}_{ij} \tag{7.20}$$

of the coordinate vectors of the particles i and j:

$$V_{ij} = V_{ij}(\mathbf{r}_{ji}), \tag{7.21}$$

see (3.39) and Figure 7.1. If the internal forces obey the strong form (7.13) of Newton's third law, V_{ij} depends on the distance

$$r_{ji} = r_{ij} = |\mathbf{r}_{ij}| = \sqrt{(x_i - x_j)^2 + (y_i - y_j)^2 + (z_i - z_j)^2}$$

only. Using the identity

$$\nabla_i r_{ji} = \frac{\mathbf{r}_i - \mathbf{r}_j}{r_{ji}} = \frac{\mathbf{r}_{ji}}{r_{ji}} = -\frac{\mathbf{r}_{ij}}{r_{ij}} = -\nabla_j r_{ij} \tag{7.22}$$

for the gradient of r_{ij} we see that (7.13) is satisfied by the forces

$$\mathbf{F}_{ji} = -\nabla_i V_{ij}(r_{ij}) = -\frac{dV_{ij}}{dr_{ij}}\nabla_i r_{ij}$$

$$= -\frac{dV_{ij}}{dr_{ij}}\frac{\mathbf{r}_{ji}}{r_{ij}} = -\left(-\frac{dV_{ij}}{dr_{ij}}\frac{\mathbf{r}_{ij}}{r_{ij}}\right) = -\mathbf{F}_{ij} \tag{7.23}$$

[1]It is necessary although somewhat cumbersome to attach indices to the interaction potential energy function. V_{ij} is not only specified by the position vectors $\mathbf{r}_i$ and $\mathbf{r}_j$, it may also depend on other variables, for example, on the masses m_i and m_j.

Assuming that also the external forces are conservative, we have

$$\mathbf{F}_i^{(e)} = -\nabla_i V_i^{(e)}, \qquad (7.24)$$

and the potential energy

$$V = \sum_{i=1}^{N} V_i^{(e)} + \frac{1}{2} \sum_{\substack{i=1 \\ }}^{N} \sum_{\substack{j=1 \\ i \neq j}}^{N} V_{ij} \qquad (7.25)$$

is related to the force $\mathbf{F}_i$ on the ith particle by

$$\mathbf{F}_i = -\nabla_i V \qquad (7.26)$$

7.4 TORQUE AND ANGULAR MOMENTUM

The total torque $\mathbf{N}$ on a system of particles relative to the origin of the coordinate system is defined by[2]

$$\mathbf{N} = \sum_{i=1}^{N} \mathbf{N}_i = \sum_{i=1}^{N} \mathbf{r}_i \times \mathbf{F}_i = \sum_{i=1}^{N} \mathbf{r}_i \times \left(\mathbf{F}_i^{(e)} + \mathbf{F}_i^{(i)} \right) \qquad (7.27)$$

It consists of the total external torque

$$\mathbf{N}^{(e)} = \sum_{i=1}^{N} \mathbf{r}_i \times \mathbf{F}_i^{(e)} \qquad (7.28)$$

produced by the external forces, and the total internal torque

$$\mathbf{N}^{(i)} = \sum_{i=1}^{N} \mathbf{r}_i \times \mathbf{F}_i^{(i)} \qquad (7.29)$$

caused by the internal forces. In contrast to the total internal force, the total internal torque does not automatically vanish when the internal forces obey the weak form of Newton's third law as the following calculation shows:

$$\mathbf{N}^{(i)} = \sum_{i=1}^{N} \mathbf{r}_i \times \mathbf{F}_i^{(i)} = \sum_{i=1}^{N} \sum_{\substack{j=1 \\ j \neq i}}^{N} \mathbf{r}_i \times \mathbf{F}_{ji}$$

$$= \frac{1}{2} \sum_{i=1}^{N} \sum_{\substack{j=1 \\ j \neq 1}}^{N} \left(\mathbf{r}_i \times \mathbf{F}_{ji} + \mathbf{r}_i \times \mathbf{F}_{ji} \right)$$

$$= \frac{1}{2} \sum_{i=1}^{N} \sum_{\substack{j=1 \\ j \neq 1}}^{N} \left(\mathbf{r}_i \times \mathbf{F}_{ji} - \mathbf{r}_i \times \mathbf{F}_{ij} \right) \qquad (7.30)$$

[2]Torque $\mathbf{N}$ should not be confused with the number of particles N.

We have used (7.12) for the last step. Interchanging the summation indices in the second term of (7.30) we obtain

$$\mathbf{N}^{(i)} = \frac{1}{2} \sum_{\substack{i=1}}^{N} \sum_{\substack{j=1 \\ j \neq i}}^{N} \left(\mathbf{r}_i \times \mathbf{F}_{ji} - \mathbf{r}_j \times \mathbf{F}_{ji} \right) = \frac{1}{2} \sum_{\substack{i=1}}^{N} \sum_{\substack{j=1 \\ j \neq i}}^{N} \left(\mathbf{r}_i - \mathbf{r}_j \right) \times \mathbf{F}_{ji}$$

$$= \frac{1}{2} \sum_{\substack{i=1}}^{N} \sum_{\substack{j=1 \\ j \neq i}}^{N} \mathbf{r}_{ji} \times \mathbf{F}_{ji} \tag{7.31}$$

which is not zero in general. Only if the forces $\mathbf{F}_{ji}$ satisfy the strong form (7.13) of Newton's third law may we conclude that $\mathbf{N}^{(i)} = 0$. Then the total torque becomes equal to the total external torque:

$$\mathbf{N} = \mathbf{N}^{(e)} \tag{7.32}$$

The total angular momentum of a system of particles relative to the origin of the coordinate system is given by

$$\mathbf{L} = \sum_{i=1}^{N} \mathbf{l}_i = \sum_{i=1}^{N} \mathbf{r}_i \times \mathbf{p}_i \tag{7.33}$$

Taking the time derivative of (7.33) we obtain

$$\dot{\mathbf{L}} = \sum_{i=1}^{N} \dot{\mathbf{r}}_i \times \mathbf{p}_i + \sum_{i=1}^{N} \mathbf{r}_i \times \dot{\mathbf{p}}_i = \sum_{i=1}^{N} \mathbf{r}_i \times \mathbf{F}_i$$

since $\dot{\mathbf{r}}_i \times \mathbf{p}_i = 0$. The term on the right-hand side of the last equation is the total torque $\mathbf{N}$. Thus

$$\dot{\mathbf{L}} = \mathbf{N} \tag{7.34}$$

For the reasons stated above both the external and the internal torque contribute to the change of the angular momentum of the system. If the conditions (7.32) are met, Eq. 7.34 simplifies to

$$\dot{\mathbf{L}} = \mathbf{N}^{(e)} \tag{7.35}$$

If also $\mathbf{N}^{(e)} = 0$, integration of (7.35) leads to

$$\mathbf{L} = \text{constant} \tag{7.36}$$

which is the theorem of angular momentum conservation.

7.5   CENTER OF MASS COORDINATE SYSTEM

For reasons to be stated shortly, it is convenient to introduce new coordinates relative to the center of mass. These center of mass coordinates (CM coordinates) are defined by

$$\mathbf{r}'_i = \mathbf{r}_i - \mathbf{R} \tag{7.37}$$

(see Figure 7.2). Trivially, we also have

$$\mathbf{r}_i = \mathbf{r}'_i + \mathbf{R}, \tag{7.38}$$

$$\dot{\mathbf{r}}_i = \dot{\mathbf{r}}'_i + \dot{\mathbf{R}} \tag{7.39}$$

and

$$\ddot{\mathbf{r}}_i = \ddot{\mathbf{r}}'_i + \ddot{\mathbf{R}} \tag{7.40}$$

The $\mathbf{r}'_i$s are not all independent of each other because

$$\sum_{i=1}^{N} m_i \mathbf{r}'_i = \sum_{i=1}^{N} m_i \mathbf{r}_i - \sum_{i=1}^{N} m_i \mathbf{R} = M\mathbf{R} - M\mathbf{R} = 0 \tag{7.41}$$

and, hence,

$$\sum_{i=1}^{N} \mathbf{p}'_i = \sum_{i=1}^{N} m_i \dot{\mathbf{r}}'_i = 0 \tag{7.42}$$

The last equation states that the total linear momentum of a system relative to the center of mass vanishes. This fact makes the CM coordinates especially useful for the description of systems for which $\sum_{i=1}^{N} \mathbf{F}_i^{(e)} = 0$, as is the case in collision processes.

Because

$$\mathbf{r}'_{ji} = \mathbf{r}'_i - \mathbf{r}'_j = (\mathbf{r}_i - \mathbf{R}) - (\mathbf{r}_j - \mathbf{R}) = \mathbf{r}_i - \mathbf{r}_j = \mathbf{r}_{ji}$$

both the potential energies $V_{ij}(\mathbf{r}_{ij})$ and the forces $\mathbf{F}_{ij}$ remain invariant under the transformation (7.37). However, if we replace $\mathbf{r}_i$ and $\dot{\mathbf{r}}_i$ by the right sides of (7.38)

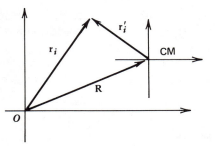

Figure 7.2 Center of mass coordinate system.

and (7.39), respectively, we obtain for the total angular momentum the expression

$$\mathbf{L} = \sum_{i=1}^{N} \mathbf{r}_i \times m_i \dot{\mathbf{r}}_i = \sum_{i=1}^{N} (\mathbf{r}_i' + \mathbf{R}) \times (m_i \dot{\mathbf{r}}_i' + m_i \dot{\mathbf{R}})$$

$$= \sum_{i=1}^{N} \mathbf{r}_i' \times m_i \dot{\mathbf{r}}_i' + \sum_{i=1}^{N} \mathbf{r}_i' \times m_i \dot{\mathbf{R}} + \sum_{i=1}^{N} \mathbf{R} \times m_i \dot{\mathbf{r}}_i' + \sum_{i=1}^{N} \mathbf{R} \times m_i \dot{\mathbf{R}}$$

The second and the third terms vanish because of the identities (7.41) and (7.42). The first term is the total angular momentum $\mathbf{L}'$ relative to the center of mass. The last term is the angular momentum of the total mass, thought of as being concentrated at the center of mass, relative to the origin of the coordinate system. Thus,

$$\mathbf{L} = \sum_{i=1}^{N} \mathbf{r}_i' \times m_i \dot{\mathbf{r}}_i' + \mathbf{R} \times \mathbf{P} = \mathbf{L}' + \mathbf{R} \times \mathbf{P} \tag{7.43}$$

A similar calculation for the kinetic energy leads to the result

$$T = \sum_{i=1}^{N} \frac{1}{2} m_i v_i^2 = \sum_{i=1}^{N} \frac{1}{2} m_i v_i'^2 + \frac{1}{2} M v_{CM}^2 = T' + \frac{1}{2} M v_{CM}^2 \tag{7.44}$$

(see Problem 7.2).

7.6 TWO-PARTICLE SYSTEMS

We now apply the formalism developed in the preceding sections to two-particle systems. As already stated, the equations of motion

$$m_1 \ddot{\mathbf{r}}_1 = \mathbf{F}_1 + \mathbf{F}_{21} = \mathbf{F}_1 - \mathbf{F}_{12} \tag{7.45a}$$

and

$$m_2 \ddot{\mathbf{r}}_2 = \mathbf{F}_2 + \mathbf{F}_{12} \tag{7.45b}$$

represent a system of coupled differential equations. In order to obtain two independent equations, we apply the center of mass coordinate transformation (7.38) to Eqs. 7.45a and 7.45b, which yield

$$m_1 \ddot{\mathbf{r}}_1' + m_1 \ddot{\mathbf{R}} = \mathbf{F}_1 - \mathbf{F}_{12} \tag{7.46a}$$

and

$$m_2 \ddot{\mathbf{r}}_2' + m_2 \ddot{\mathbf{R}} = \mathbf{F}_2 + \mathbf{F}_{12} \tag{7.46b}$$

Adding the last two equations, we find

$$(m_1 + m_2)\ddot{\mathbf{R}} = \mathbf{F}_1 + \mathbf{F}_2$$

or

$$M\ddot{\mathbf{R}} = \mathbf{F} \tag{7.47}$$

which is (7.16) applied to two particles. If we multiply (7.46a) and (7.46b) by $1/m_1$ and $1/m_2$, respectively, and subtract the first from the second equation, we obtain

$$\ddot{\mathbf{r}}_1' - \ddot{\mathbf{r}}_1' = \frac{1}{m_2}\mathbf{F}_2 - \frac{1}{m_1}\mathbf{F}_1 + \left(\frac{1}{m_1} + \frac{1}{m_2}\right)\mathbf{F}_{12} \tag{7.48}$$

With the relative coordinate vector

$$\mathbf{r} = \mathbf{r}_{12} = \mathbf{r}_2' - \mathbf{r}_1' = \mathbf{r}_2 - \mathbf{r}_1, \tag{7.49}$$

(see Figure 7.3), and the reduced mass μ defined by

$$\frac{1}{\mu} = \frac{1}{m_1} + \frac{1}{m_2}, \tag{7.50}$$

(7.48) goes over into

$$\mu\ddot{\mathbf{r}} = \mu\left(\frac{1}{m_2}\mathbf{F}_2 - \frac{1}{m_1}\mathbf{F}_1\right) + \mathbf{F}_{12} \tag{7.51}$$

Equations 7.47 and 7.51 are the desired decoupled equations of motion. They represent two linearly independent single-particle equations. Equation 7.47 relates the total force to the total mass M of the system and to the center of mass acceleration. (This fact was already established from the general discussion of the motion of systems of particles). Equation 7.51 relates the force $\mathbf{F}_{12}$, in addition to

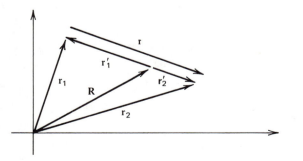

Figure 7.3 Relations between position vectors.

$\mathbf{F}_1$ and $\mathbf{F}_2$, to the reduced mass and the acceleration of the relative position vector. We may think of two fictitious mass points, one of mass M at the center of mass, the other of mass μ at $\mathbf{r}$ relative to the center of mass.

In situations where the accelerations of both mass points caused by the external forces are equal to each other,

$$\frac{\mathbf{F}_1^{(e)}}{m_1} = \mathbf{a}_1^{(e)} = \mathbf{a}_2^{(e)} = \frac{\mathbf{F}_2^{(e)}}{m_2},$$

(7.51) simplifies to

$$\mu\ddot{\mathbf{r}} = \mathbf{F}_{12} \tag{7.52}$$

Trivially, this equation also holds if there are no external forces, in which case we also have

$$M\ddot{\mathbf{R}} = 0$$

Without loss of generality we may then set $\mathbf{R} = 0$.

If $\mathbf{R}(t)$ and $\mathbf{r}(t)$ are known from (7.47) and (7.51) or (7.52), the coordinate vectors $\mathbf{r}_1(t)$ and $\mathbf{r}_2(t)$ of the mass points can be obtained from

$$\mathbf{r}_1 = \mathbf{R} - \frac{m_2}{m_1 + m_2}\mathbf{r} \tag{7.53a}$$

and

$$\mathbf{r}_2 = \mathbf{R} + \frac{m_1}{m_1 + m_2}\mathbf{r} \tag{7.53b}$$

(see Problem 7.3). Thus, provided that Eqs. 7.47 and 7.51 are solvable, the two-particle problem can always be solved.

It is instructive to picture the motion of the four particles m_1, m_2, M, and μ. The center of mass follows a path that is determined by the external forces (see Figure 7.4). Relative to the coordinate system fixed in space, the mass points m_1 and m_2 trace out complicated orbits. Relative to the center of mass frame, the motion of the particles is much simpler. Because all four mass points lie on a

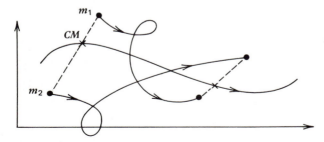

Figure 7.4 Orbits of two interacting particles in a fixed reference frame.

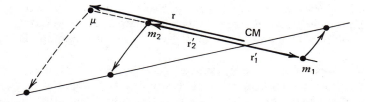

Figure 7.5 Orbits of two interacting particles in the center of mass frame.

straight line, m_1, m_2, and μ move about the center of mass with the same angular velocity (see Figure 7.5).

It is often the case that the mass of one particle is much larger than the mass of the other particle, say, for example, $m_1 \gg m_2$ [sun-planet, earth-satellite, proton-electron (hydrogen atom)]. We then have

$$M = m_1 + m_2 \approx m \quad \text{and} \quad \mu = \frac{m_1 m_2}{m_1 + m_2} \approx m_2$$

The reduced mass of the hydrogen atom, for example, is only about 0.05 percent smaller than the electron mass, but the difference still has an observable effect on the spectrum. If $m_1 \gg m_2$, the radius vectors $\mathbf{r}_1$ and $\mathbf{r}_2$ become

$$\mathbf{r}_1 \approx \mathbf{R} \quad \text{and} \quad \mathbf{r}_2 \approx \mathbf{R} + \mathbf{r},$$

which can easily be verified by using (7.53a) and (7.53b). The lighter particle m_2 moves about m_1, which is located at the center of mass to the first approximation.

Example 7.1

Let m_1 be the mass of the sun and m_2 be the mass of a planet. Because $m_1 m_2 = \mu M$, the internal (gravitational) force on the planet is

$$\mathbf{F}_{12} = -G\frac{m_1 m_2}{r^2}\frac{\mathbf{r}}{r} = -G\frac{\mu M}{r^2}\frac{\mathbf{r}}{r}$$

In the absence of external forces, Eq. 7.52 holds such that

$$\mu\ddot{\mathbf{r}} = -G\frac{\mu M}{r^2}\frac{\mathbf{r}}{r} \tag{7.54}$$

and the orbit of μ is a conic section (see Section 5.6). ∎

7.7 COLLISION OF TWO PARTICLES

A collision is an interaction of two or more particles in relative motion with forces that act during a short time interval or, equivalently, that are of short range. Forces of basically long range may be present, but we assume their effect on the motion to be negligible. For example, the long-range electrostatic force between an α particle and a nucleus is practically zero at distances larger than the radius of the atom because of the screening effect of the electrons surrounding the nucleus.

In Section 6.8 we investigated some dynamical consequences of the special form of the force law of the interaction, namely, cross sections. Here we focus our attention on the overall kinematic behavior of the particles before and after the collision without taking into account the special form of the force of interaction. The formulas that will be derived in this and the following section are therefore applicable to collision problems whether they are treated in (nonrelativistic) classical mechanics or in quantum mechanics.

We restrict ourselves to two-particle collisions. In dilute gases, three-particle collisions between the molecule occur with much less frequency than two-particle collisions. We assume without loss of generality that external forces are absent. Then, the particles move freely before and after the collision, and their momenta and energies assume constant values.

Before the collision:

$$\mathbf{p}_i = m_i \mathbf{v}_i = \text{constant}, \quad E_i = T_i = \tfrac{1}{2} m_i v_i^2 = \text{constant}$$

After the collision:

$$\bar{\mathbf{p}}_i = m_i \bar{\mathbf{v}}_i = \text{constant}, \quad \bar{E}_i = \bar{T}_i = \tfrac{1}{2} m_i \bar{v}_i^2 = \text{constant}.$$

Because the equations of motion

$$\dot{\mathbf{p}}_1 = \mathbf{F}_{21} \quad \text{and} \quad \dot{\mathbf{p}}_2 = \mathbf{F}_{12}$$

involve only internal forces, the total momentum of the system is conserved:

$$\mathbf{p}_1 + \mathbf{p}_2 = \bar{\mathbf{p}}_1 + \bar{\mathbf{p}}_2 \tag{7.55}$$

The change of the momenta of the two particles during the collision is equal to the impulse delivered by the force of interaction during the collision time $\Delta t = t_1 - t_2$:

$$\bar{\mathbf{p}}_1 - \mathbf{p}_1 = -(\bar{\mathbf{p}}_2 - \mathbf{p}_2) = \int_{t_1}^{t_2} \mathbf{F}_{21} \, dt$$

The change of the total energy produced by the force of interaction during the collision is called the Q-value

$$Q = (T_1 + T_2) - (\bar{T}_1 + \bar{T}_2) \tag{7.56}$$

Depending on the nature of the interaction one distinguishes between the following cases:

(a) Exoergic collision: $Q < 0$, kinetic energy is gained (e.g., explosions)
(b) Elastic collision: $Q = 0$, kinetic energy is conserved.
(c) Endoergic collision: $Q > 0$, kinetic energy is lost (e.g., inelastic collision)

In one dimension the momentum and energy equations (7.55) and (7.56) reduce to

$$m_1 v_1 + m_2 v_2 = m_1 \bar{v}_1 + m_2 \bar{v}_2 \tag{7.57}$$

and

$$\tfrac{1}{2} m_1 v_1^2 + \tfrac{1}{2} m_2 v_2^2 = \tfrac{1}{2} m_1 \bar{v}_1^2 + \tfrac{1}{2} m_2 \bar{v}_2^2 + Q \tag{7.58}$$

It is convenient to introduce the ratio ε of the relative speed of separation to the relative speed of approach

$$\varepsilon = \frac{\bar{v}_2 - \bar{v}_1}{v_1 - v_2} \tag{7.59}$$

which is called the coefficient of restitution. Equations 7.57 and 7.59 may be solved for the final velocities in terms of the initial velocities. The result is

$$\bar{v}_1 = \frac{m_1 - \varepsilon m_2}{M} v_1 + \frac{(1 + \varepsilon) m_2}{M} v_2 \tag{7.60}$$

$$\bar{v}_2 = \frac{(1 + \varepsilon) m_1}{M} v_1 + \frac{-\varepsilon m_1 + m_2}{M} v_2 \tag{7.61}$$

where $M = m_1 + m_2$ is the total mass. Substituting (7.60) and (7.61) into (7.58), we obtain after a lengthy calculation

$$Q = \tfrac{1}{2} \mu (v_1 - v_2)^2 (1 - \varepsilon^2) \tag{7.62}$$

where $\mu = m_1 m_2 / M$ is the reduced mass.

Equation 7.62 enables us to analyze quickly the three types of collisions.

(a) Exoergic collision: $Q < 0$ implies $\varepsilon^2 > 1$. The relative speed of separation is larger than the relative speed of approach.
(b) Elastic collision: $Q = 0$ implies $\varepsilon^2 = 1$. The relative speed of separation is equal to the relative speed of approach:

$$v_1 - v_2 = \bar{v}_2 - \bar{v}_1$$

Then (7.60) and (7.61) go over into

$$\bar{v}_1 = \frac{m_1 - m_2}{M} v_1 + \frac{2 m_2}{M} v_2,$$

$$\bar{v}_2 = \frac{2 m_1}{M} v_1 + \frac{-m_1 + m_2}{M} v_2$$

If furthermore $m_1 = m_2$, the particles simply exchange their velocities during the collision: $v_1 = v_2$, $v_2 = v_1$.

(c) Endoergic collision: $Q > 0$ implies $\varepsilon^2 < 1$. For $\varepsilon = 0$, Q reaches the maximum value

$$Q = \tfrac{1}{2}\mu(v_1 - v_2)^2$$

In this case of a perfectly inelastic collision, the final velocities of the particles are equal to each other, $\bar{v}_1 = \bar{v}_2$.

Collisions of particles in two and three dimensions can be treated in a similar manner. Some aspects of collisions in two dimensions are discussed in the next section.

7.8 TRANSFORMATIONS BETWEEN THE LABORATORY AND THE CENTER OF MASS COORDINATE SYSTEMS

Two reference frames are of particular importance for the description of collision processes. In the laboratory frame the velocities and momenta of the particles are measured relative to the fixed laboratory. This frame is convenient for the experimental physicist, especially when one of the two particles, the target particle, is initially at rest: $\mathbf{v}_2 = 0$. This holds, for example, for the nucleus in Rutherford's scattering of α particles. In recent years, however, colliding beam experiments have become more and more important because the needed collision energies cannot be reached with fixed targets in any feasible accelerator (see Problem 7.17). In the center of mass frame (CM frame for short) the velocities are measured relative to the center of mass (see Section 7.5). In the CM frame the total momentum vanishes (see Eq. 7.42), which simplifies the calculations:

$$\mathbf{p}_1' + \mathbf{p}_2' = 0, \quad \bar{\mathbf{p}}_1' + \bar{\mathbf{p}}_2' = 0 \tag{7.63}$$

For elastic collisions we also have

$$p_1' = \bar{p}_1', \tag{7.64}$$

(see Problem 7.10), such that

$$p_1' = p_2' = \bar{p}_1' = \bar{p}_2' \tag{7.65}$$

or

$$m_1 v_1' = m_2 v_2' = m_1 \bar{v}_1' = m_2 \bar{v}_2' \tag{7.66}$$

This equation shows that the endpoints of the four momentum vectors $\mathbf{p}_1'$, $\mathbf{p}_2'$, $\bar{\mathbf{p}}_1'$, and $\bar{\mathbf{p}}_2'$ lie on a circle (see Figure 7.6).

We assume now that the target particle is initially at rest in the laboratory frame. Our goal is to find a relation between the scattering angles θ and θ' in the

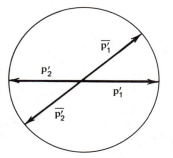

Figure 7.6 Momentum vectors in the center of mass frame (elastic collision).

laboratory and CM frames, respectively. We first apply the transformation equation (7.39) to the particles' velocities before and after the collision. The vectors are drawn in Figures 7.7*a–d* for clarification.
Before the collision:

$$\mathbf{v}_1 = \mathbf{v}_1' + \mathbf{V}, \quad 0 = \mathbf{v}_2 = \mathbf{v}_2' + \mathbf{V} \tag{7.67a, b}$$

After the collision:

$$\bar{\mathbf{v}}_1 = \bar{\mathbf{v}}_1' + \mathbf{V}, \quad \bar{\mathbf{v}}_2 = \bar{\mathbf{v}}_2' + \mathbf{V} \tag{7.67c, d}$$

Because the target particle is initially at rest, it is obvious that the relative velocity of approach is equal to the initial velocity $\mathbf{v}_1$ of the incident particle, $\mathbf{v}_1 - \mathbf{v}_2 = \mathbf{v}_1$. From Figure 7.7*c* we read off the relations

$$\bar{v}_1 \cos \theta = V + \bar{v}_1' \cos \theta' \tag{7.68}$$

and

$$\bar{v}_1 \sin \theta = \bar{v}_1' \sin \theta' \tag{7.69}$$

Dividing (7.69) by (7.68), we obtain

$$\tan \theta = \frac{\bar{v}_1' \sin \theta'}{V + \bar{v}_1' \cos \theta'} = \frac{\sin \theta'}{\dfrac{V}{\bar{v}_1'} + \cos \theta'} \tag{7.70}$$

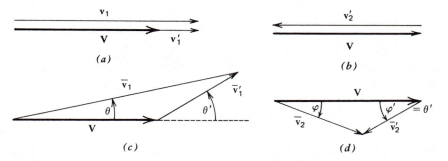

Figure 7.7 Velocity vectors for two-particle collisions.

For elastic collisions the term $V/\bar{v}_1'$ is equal to

$$\frac{V}{\bar{v}_1'} = \frac{v_2'}{\bar{v}_1'} = \frac{m_1}{m_2}$$

because of Eqs. 7.67b and 7.66, such that finally

$$\tan\theta = \frac{\sin\theta'}{\dfrac{m_1}{m_2} + \cos\theta'} \tag{7.71}$$

If further $m_1 = m_2$ (e.g., proton-proton scattering, collision of two billiard balls), (7.71) may be written

$$\tan\theta = \frac{\sin\theta'}{1 + \cos\theta'} = \frac{\sin 2\dfrac{\theta'}{2}}{1 + \cos 2\dfrac{\theta'}{2}} = \frac{2\sin\dfrac{\theta'}{2}\cos\dfrac{\theta'}{2}}{1 + \cos^2\dfrac{\theta'}{2} - \sin^2\dfrac{\theta'}{2}} = \tan\frac{\theta'}{2}$$

from which follows

$$\theta = \frac{\theta'}{2} \tag{7.72}$$

Because $\theta_{max}' = \pi$, we obtain $\theta_{max} = \pi/2$. Thus, backscattering is impossible if $m_1 = m_2$.

The connection between the collision cross sections in the laboratory and the CM frames is established as follows. Because the impact parameter has the same value in both the laboratory frame and the CM frame, (6.65) implies that

$$2\pi\sigma(\theta)\sin\theta\, d\theta = 2\pi s\, ds = 2\pi\sigma'(\theta')\sin\theta'\, d\theta'$$

and, hence,

$$\sigma'(\theta') = \sigma(\theta)\frac{\sin\theta\, d\theta}{\sin\theta'\, d\theta'} = \sigma(\theta)\frac{d\cos\theta}{d\cos\theta'}$$

For example, if the conditions of Eq. 7.72 are met, we find

$$\sigma'(\theta') = \sigma'(2\theta) = \frac{\sigma(\theta)}{4\cos\theta}$$

In most practical situations the calculations are much more complicated than those discussed here.

PROBLEMS

7.1 Show whether or not a force of the form $\mathbf{F}_{ij} = a(\mathbf{r}_i - \mathbf{r}_j) \times (\mathbf{v}_i - \mathbf{v}_j)$ satisfies Newton's third law.

7.2 Derive Eq. 7.44.

7.3 Derive Eqs. 7.53a and 7.53b.

7.4 Express (a) the kinetic energy, and (b) the total angular momentum vector of a two-particle system in terms of the quantities $\mathbf{R}$, $\mathbf{r}$, $\dot{\mathbf{R}}$, $\dot{\mathbf{r}}$, M, and μ.

7.5 Find the exact form of Kepler's third law which includes the mass of the planet.

7.6 Two objects of masses m_1 and m_2, respectively, are connected by a spring of force constant k. The system is free to oscillate without friction. Find the frequency of oscillation.

7.7 Derive Eq. 7.62.

7.8 A tennis ball is dropped from height h and rebounds from the surface of the tennis court with coefficient of restitution ε.

(a) After what time will the ball come to rest?
(b) What is the total distance traveled?

7.9 The following collisions are one-dimensional and perfectly inelastic.

(a) Particle 1 of mass m_1 and speed v collides with particle 2 of mass m_2, initially at rest. Thereafter, both particles collide with particle 3 of mass m_3, initially at rest.
(b) Particle 1 of mass m_1 and speed v collides with another particle of mass $m_2 + m_3$, initially at rest. Calculate the total energy losses for both cases. Compare the results and explain the outcome.

7.10 Derive Eq. 7.64.

7.11 Is it conceivable that forces exist such that the momentum vectors $\mathbf{p}_1$ and $\mathbf{p}_2$ of two particles before the collision and the momentum vectors $\bar{\mathbf{p}}_1$ and $\bar{\mathbf{p}}_2$ after the collision do not lie in the same plane? If the answer is "yes," can you give examples of such forces? What can you say about angular momentum conservation in such a case?

7.12 A particle collides with a target particle of the same mass and initially at rest.

(a) Find the Q-value of the reaction in terms of the final speeds of the two particles and the angle ψ between the final velocities in the laboratory frame.
(b) What is the value of ψ for elastic collision?

7.13 Two particles undergo an inelastic collision in two dimensions. Compare the energy loss in the laboratory frame with that in the CM frame.

7.14 A particle of mass m and initial speed v collides with a smooth plane surface (See Figure 7.8).

(a) Find the angle of reflection β in terms of the angle of incidence α and the coefficient of restitution ε.
(b) Find the speed of the particle after the collision in terms of v, α, and ε.
(c) Find the Q-value in terms of v, α, and ε.

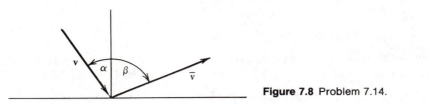

Figure 7.8 Problem 7.14.

7.15 Assume that the target particle is initially at rest. Derive a relation between the angles φ and φ' describing the directions of the motion of the target particle after an elastic collision in the laboratory and CM frames, respectively. (See Figure 7.7*d*.)

7.16 (a) A deuteron of energy 3 MeV collides elastically with a proton initially at rest. Find the maximum deflection angles θ and φ in the laboratory frame. Set $m_d = 2m_p$. Apply the result obtained in Problem 7.15.

(b) Find the energies of the particles after the collision.

7.17 The radius r of an accelerator for *colliding* beams is nearly proportional to the maximal collision energy E. If such a machine was used with a *fixed* target, the collision energy E of particles with rest energy $mc^2 = E_0 \ll E$ would decrease to $T = \sqrt{2E_0 E}$. The Large Electron-Positron Project (LEP) to be constructed under Geneva must have $r = 5$ km in order to reach $E/E_0 \approx 500\,000$ (and will consume about 10^8 W).

(a) Which collision energy T would the same accelerator achieve in a target?

(b) What radius R would be required to reach the desired E with a fixed target?

(c) Compare the needed power with that produced in a nuclear or hydro-electric plant.

chapter

8

The Formalisms of Lagrange and Hamilton

Daring ideas are like chessmen moved forward;
they may be beaten, but they may start a winning game.

GOETHE

8.1 INTRODUCTION

Single particles or systems of particles are in general subjected to three types of conditions that determine the motion of these systems in space and time. In the preceding chapters we considered only two of these conditions, namely, *dynamical conditions*, which are expressed by the forces acting on the system via the equations of motion, and *initial conditions*, which are usually expressed by the initial values of the position and the velocity coordinates of the particles and/or by some conserved dynamical quantities. Initial conditions single out the actual orbit from the infinitely many possible solutions of the equations of motion. Newton's laws are perfectly suited to investigate the behavior of systems that are subjected to these two types of conditions only. Occasionally it turned out that a particle (that was subjected to these conditions but was otherwise free to move in the three-dimensional space) described orbits in only two dimensions (e.g., under the influence of a central force), or it turned out that the orbit was closed (e.g., under the influence of the isotropic linear harmonic oscillator force). In these cases, the special motion of the particles in certain subspaces of the three-dimensional space resulted from the dynamical symmetries of the force laws.

An entirely new situation arises if the system is a priori subjected to *geometrical conditions* or *constraints*. These are conditions imposed on the coordinates of the particles without reference to the force laws. For example, the bob of a spherical pendulum is allowed to move only on the surface of a sphere. Geometrical conditions may cause the particles of a system to accelerate into directions that are different from the directions of the vector sum of all internal and external forces acting on the particles when no constraints are present. Additional forces, the so-called forces of constraints, appear when the system is subjected to constraints.

If all forces of constraints, together with the other forces acting on the system were known, Newton's formalism (which is expressed in terms of the dynamical and the initial conditions only) would be sufficient to determine the behavior of the system in space and time. In practically all cases, however, the constraints are known, whereas the forces of constraints are unknown, and in order to apply Newton's equations of motion, one must first calculate the forces of constraints from the equations of the constraints and from the forces impressed on the system. In certain situations, the equations of motion are also needed. (On the other hand, the geometrical conditions may be obtained from the forces of constraints along with the other information.) Fortunately, as we shall see in this chapter, such a detour is not necessary if one is interested only in the time development of the system. A straightforward and rather simple procedure exists for setting up the equations of motion that do not contain the forces of constraints. Moreover, the concept of force is eliminated entirely. It turns out that this formalism that was introduced by Lagrange is equally applicable to systems that are free from constraints.

8.2 CONSTRAINTS AND GENERALIZED COORDINATES

Let us consider a system of N particles that may be described by the $3N$ time-dependent position coordinates

$$x_1, y_1, z_1, x_2, \ldots, z_N \tag{8.1}$$

and the velocity coordinates

$$\dot{x}_1, \dot{y}_1, \dot{z}_1, \dot{x}_2, \ldots, \dot{z}_N \tag{8.2}$$

in a Cartesian coordinate system. One says that the $3N$ coordinates (8.1) are elements of a $3N$-dimensional coordinate space, and any point in this $3N$-dimensional coordinate space describes a configuration of the system in real space.

If the system is not subjected to any constraints, the particles are free to move in the 3-dimensional space, and the coordinates are all independent of each other. If, however, constraints are imposed on the system, the coordinates are not all independent of each other, or they may not be allowed to assume arbitrary values.

One distinguishes between the following types of constraints:

A constraint is called holonomic[1] if an algebraic relation exists between the coordinates. It can be expressed by an equation of the form

$$f(x_1, y_1, z_1, x_2, \ldots z_N) = 0 \tag{8.3}$$

Example 8.1

The bob of a spherical pendulum of length l is restricted to move on a sphere described by $x^2 + y^2 + z^2 = l^2$. This equation may also be written in the form

$$f(x, y, z) = x^2 + y^2 + z^2 - l^2 = 0$$

which shows that the constraint is holonomic. ∎

---

[1]Holos (Greek): whole.

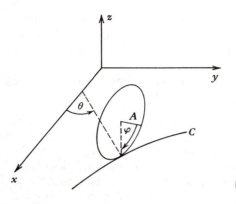

Figure 8.1 The rolling disc.

A constraint is called nonholonomic if it cannot be written in the form (8.3). The most common nonholonomic constraints are expressed in the form of inequalities, or they involve differentials of the coordinates.

Example 8.2

The motion of a gas molecule in a cubic container of side L is restricted by the nonholonomic conditions $0 \leqslant x \leqslant L$, $0 \leqslant y \leqslant L$, and $0 \leqslant z \leqslant L$. ∎

Example 8.3

A coin of radius A rolls in vertical position on the x-y plane (see Figure 8.1). Let θ be the angle between the x-axis and the normal of the coin, and let φ be the angle of rotation of the coin. Because the speed of the coin is $v = A\dot{\varphi}$, the velocity coordinates are $v_x = \dot{x} = v \cos \theta$ and $v_y = \dot{y} = v \sin \theta$, from which follow the nonholonomic conditions

$$dx = A \cos \theta \, d\varphi \quad \text{and} \quad dy = A \sin \theta \, d\varphi$$ ∎

Another classification of the constraints is made according to whether or not a constraint is time dependent. Constraints of the first class are called rheonomous,[2] constraints of the second class scleronomous.[3]

Example 8.4

The molecules of a gas in a container of changing volume are subjected to rheonomous nonholonomic constraints. ∎

A general method for solving problems involving nonholonomic constraints does not exist. Each case must be investigated separately. If the nonholonomic constraints are described by differential equations, then these differential equa-

[2]Rheos (Greek): current.
[3]Skleros (Greek): hard.

tions must be solved simultaneously with the system of the equations of motion. It is very fortunate that most problems of classical mechanics as well as of modern physics involve only holonomic constraints. Therefore, we shall concentrate on this case only.

We consider now a system which is subjected to c independent holonomic constraints. Then there exist c independent equations of the form (8.3):

$$f_k(x_1, y_1, z_1, x_2, \ldots z_N) = 0, \quad k = 1, 2, \ldots c \tag{8.4}$$

Each equation may be solved for any one of the coordinates in terms of the remaining $3N - 1$ coordinates (at least in principle in a small neighborhood about a point in the three-dimensional coordinate space). It is then possible to eliminate any c of the $3N$ original coordinates, thus reducing the number of independent coordinates to

$$f = 3N - c \tag{8.5}$$

This f is called the number of macroscopic degrees of freedom of the system, or degrees of freedom for short. It is larger than the number of the "microscopic" degrees of freedom that we obtain when nonholonomic constraints are imposed on the system.

Example 8.5

The coin of Example 8.3 may be rolled to any point in the x-y plane, and it may be brought to that point in any orientation defined by the angles θ and φ by choosing an appropriate path. The coordinates necessary to describe the system completely are x, y, θ, and φ; hence, the number of macroscopic degrees of freedom is four. But because the coin is not allowed to slide, the motion along the axis of rotation of the coin, and a change of the angle φ at any fixed point (x, y) are forbidden. The coin can move only along the one-dimensional curve C or rotate about the vertical axis. Therefore, the number of microscopic degrees of freedom of the system is two. ∎

Any set of f independent coordinates that (macroscopically) completely specifies the system is called a set of generalized coordinates. We denote these generalized coordinates by

$$q_1, q_2, \ldots q_f \tag{8.6}$$

The time derivatives

$$\dot{q}_1, \dot{q}_2, \ldots \dot{q}_f \tag{8.7}$$

are called generalized velocities. In mathematical terms, the set of equations (8.4) describes a projection from the $3N$-dimensional coordinate space to an f-dimensional subspace, which is called the configuration space. The configuration of the system is uniquely described by a point in that f-dimensional configuration space. In the special case of no holonomic constraints, the number of degrees of freedom is $f = 3N$, and any $3N$ independent coordinates are already generalized coordinates. The generalized coordinates q_j need not necessarily have the dimension of a

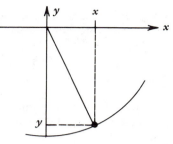

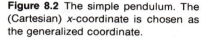

Figure 8.2 The simple pendulum. The (Cartesian) x-coordinate is chosen as the generalized coordinate.

length, a fact that has important consequences when generalized forces and generalized momenta are introduced.

Example 8.6

For a single mass point that is free to move in the three-dimensional space, either the Cartesian coordinates (x, y, z) or the circular cylindrical coordinates (ρ, φ, z) or the spherical polar coordinates (r, θ, φ) or any other set of coordinates may be chosen as the generalized coordinates (q_1, q_2, q_3). ∎

Example 8.7

The bob of a simple pendulum moving in a plane is subjected to the holonomic constraint $x^2 + y^2 - l^2 = 0$. The number of degrees of freedom is one. The x-coordinate (see Figure 8.2), may be selected as the generalized coordinate q because $y = y(x) = (-)\sqrt{l^2 - x^2}$. However, the equation of motion would turn out to be very awkward (see Problem 8.2):

$$m\ddot{x} = -mg\frac{x}{l^2}\sqrt{l^2 - x^2} - m\frac{x}{l^2 - x^2}\dot{x}^2 \tag{8.8}$$

Another possible and more natural choice of the generalized coordinate is the angle θ, because $r = l = $ constant (see Figure 8.3). By (5.58) the equation of motion reads simply

$$ml^2\ddot{\theta} = -mgl\sin\theta \qquad\qquad ∎$$

This example shows that one should always seek "natural" coordinates which relate directly to the symmetries of the system, and in terms of which the

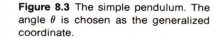

Figure 8.3 The simple pendulum. The angle θ is chosen as the generalized coordinate.

equations of motion take on a much simpler form than in other coordinates. Experience and intuition may be required to find such "natural" coordinates for a given system. Because most examples are usually formulated in these "natural" coordinates, students who study this subject for the first time are often misled into believing that there is only one set of generalized coordinates for each system.

Let us summarize: We have seen that a system of N particles which is subjected to c holonomic constraints is completely specified by a set of $f = 3N - c$ generalized coordinates. We therefore expect that it should be possible to find equations of motion for the system in terms of these generalized coordinates. In fact, these equations exist, and they are named after the discoverer Lagrange's equations. A little more groundwork must be done before we can derive Lagrange's equations.

8.3 IDENTITIES INVOLVING GENERALIZED COORDINATES

We assume that a set of generalized coordinates q_j, $j = 1, 2, \ldots f$, has been selected for a given system. The Cartesian coordinates x_i, y_i, and z_i of the position vector $\mathbf{r}_i$, $i = 1, 2, \ldots N$, may then be considered as functions of the q_j. Using vector notation we have

$$\mathbf{r}_i = \mathbf{r}_i(q_1, \ldots q_f) = \mathbf{r}_i(q_1(t), \ldots q_f(t)), \quad i = 1, \ldots N \tag{8.9}$$

The $\mathbf{r}_i$ depend on the time implicitly via the q_j. If the system is subjected to rheonomous constraints, the $\mathbf{r}_i$ also depend on the time explicitly, in which case we have

$$\mathbf{r}_i = \mathbf{r}_i(q_1(t), \ldots q_f(t), t) \tag{8.10}$$

The velocity vector is then equal to

$$\frac{d\mathbf{r}_i}{dt} = \dot{\mathbf{r}}_i = \sum_{j=1}^{f} \frac{\partial \mathbf{r}_i}{\partial q_j} \dot{q}_j + \frac{\partial \mathbf{r}_i}{\partial t} \tag{8.11}$$

Two identities can be derived from (8.11). Performing partial differentiation with respect to the generalized velocity $\dot{q}_k$ we find

$$\frac{\partial \dot{\mathbf{r}}_i}{\partial \dot{q}_k} = \sum_{j=1}^{f} \frac{\partial \mathbf{r}_i}{\partial q_j} \frac{\partial \dot{q}_j}{\partial \dot{q}_k} + \frac{\partial}{\partial \dot{q}_k} \left(\frac{\partial \mathbf{r}_i}{\partial t} \right)$$

$$= \sum_{j=1}^{f} \frac{\partial \mathbf{r}_i}{\partial q_j} \delta_{jk}$$

$$= \frac{\partial \mathbf{r}_i}{\partial q_k}, \tag{8.12}$$

because $\partial \mathbf{r}_i / \partial t$ does not depend on $\dot{q}_k$. Taking the partial derivative of $\dot{\mathbf{r}}_i$ with respect to q_k and interchanging the partial differentiations, we obtain

$$\frac{\partial \dot{\mathbf{r}}_i}{\partial q_k} = \sum_{j=1}^{f} \frac{\partial}{\partial q_k}\left(\frac{\partial \mathbf{r}_i}{\partial q_j}\right)\dot{q}_j + \frac{\partial}{\partial q_k}\left(\frac{\partial \mathbf{r}_i}{\partial t}\right)$$

$$= \sum_{j=1}^{f} \frac{\partial}{\partial q_j}\left(\frac{\partial \mathbf{r}_i}{\partial q_k}\right)\dot{q}_j + \frac{\partial}{\partial t}\left(\frac{\partial \mathbf{r}_i}{\partial q_k}\right)$$

$$= \frac{d}{dt}\left(\frac{\partial \mathbf{r}_i}{\partial q_k}\right) \tag{8.13}$$

Equation 8.13 states that the differential operators d/dt and $\partial/\partial q_k$ may be interchanged.

8.4 VIRTUAL DISPLACEMENT, VIRTUAL WORK, AND GENERALIZED FORCES

One distinguishes between two types of infinitesimal displacements of a mass point m_i. The displacement is called an actual displacement $d\mathbf{r}_i$ if the particle moves from $\mathbf{r}_i$ to $\mathbf{r}_i + d\mathbf{r}_i$ during the time interval dt, under the influence of the applied forces, and consistent with the equations of motion and consistent with the constraints. The displacement is called a virtual[4] displacement $\delta\mathbf{r}_i$ if it is consistent with only the constraints. A virtual displacement of the particle does not involve the equations of motion or the time. For example, the bob of a simple pendulum may be moved from the point (R, φ) to the point $(R, \varphi + \delta\varphi)$ (in plane polar coordinates) during any arbitrary time interval as long as the bob remains on the circle of radius R at all times. It follows then from (8.9) that any virtual displacement obeys

$$\delta\mathbf{r}_i = \sum_{j=1}^{f} \frac{\partial \mathbf{r}_i}{\partial q_j}\delta q_j \tag{8.14}$$

where δq_i is the virtual change of the generalized coordinate q_j.

The total virtual work δW is defined as the total work done by the forces acting on the system whereby each particle undergoes the virtual displacement $\delta\mathbf{r}_i$, $i = 1, 2, \ldots N$. Using Eq. 8.14 we obtain

$$\delta W = \sum_{i=1}^{N} \mathbf{F}_i \cdot \delta\mathbf{r}_i$$

$$= \sum_{i=1}^{N} \mathbf{F}_i \cdot \left(\sum_{j=1}^{f} \frac{\partial \mathbf{r}_i}{\partial q_j}\delta q_j\right)$$

$$= \sum_{j=1}^{f}\left(\sum_{i=1}^{N} \mathbf{F}_i \cdot \frac{\partial \mathbf{r}_i}{\partial q_j}\right)\delta q_j \tag{8.15}$$

[4]Virtual = being so in effect or essence, although not in actual fact or name, *Webster's New World Dictionary*, College Edition.

Introducing the generalized force Q_j associated with the generalized coordinate q_j by

$$Q_j = \sum_{i=1}^{N} \mathbf{F}_i \cdot \frac{\partial \mathbf{r}_i}{\partial q_j} \tag{8.16}$$

we find

$$\delta W = \sum_{j=1}^{f} Q_j \, \delta q_j \tag{8.17}$$

Two important differences between the forces $\mathbf{F}_i$ and the generalized forces Q_j need to be emphasized. Although $\mathbf{F}_i$ is a vector in the three-dimensional space, Q_j is no longer a vector. The virtual work always has the dimension of energy, but because the generalized coordinate q_j does not necessarily have the dimension of a length, the generalized force Q_j associated with q_j does not necessarily have the dimension of a force.

Example 8.8

A force $\mathbf{F}$ acts on a mass point that is constrained to move on a circle of radius R (see Figure 8.4). If we choose the angle φ as the generalized coordinate, we have

$$\frac{\partial x}{\partial \varphi} = \frac{\partial}{\partial \varphi}(R \cos \varphi) = -R \sin \varphi \quad \text{and} \quad \frac{\partial y}{\partial \varphi} = \frac{\partial}{\partial \varphi}(R \sin \varphi) = R \cos \varphi$$

Using (8.16) and some trigonometric identities, we obtain for the associated generalized force Q_φ

$$Q_\varphi = \mathbf{F} \cdot \frac{\partial \mathbf{r}}{\partial \varphi} = F_x \frac{\partial x}{\partial \varphi} + F_y \frac{\partial y}{\partial \varphi}$$

$$= F \cos(\varphi + \theta)(-R \sin \varphi) + F \sin(\varphi + \theta) R \cos \varphi$$

$$= RF \sin \theta = |\mathbf{r} \times \mathbf{F}| = N$$

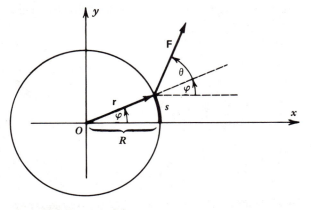

Figure 8.4 Example 8.8.

which is the torque acting on the mass point. If we had chosen the arc length $s = R\theta$ as the generalized coordinate, the associated generalized force would have been $Q_s = F \sin \theta$, which is the tangential component of the force. Note that the dimensions of Q_φ and Q_s are different, whereas the virtual work assumes the same dimension for each pair of the corresponding generalized coordinates and forces:

$$\delta W = Q_\varphi \, \delta_\varphi = (RF \sin \theta) \, \delta\varphi$$

$$= (F \sin \theta)(R \, \delta\varphi) = Q_s \, \delta s \qquad \blacksquare$$

This example is instructive for another reason. The evaluation of the generalized force using (8.16) may turn out to be very tedious in many cases. It is often more advantageous to derive the generalized force from the virtual work directly by investigating the physical meaning of Q_j. We could have obtained Q_φ and Q_s in Example 8.8 rather easily that way from δW.

We consider now a system of N mass points in static equilibrium. In Newtonian mechanics, equilibrium is expressed by the conditions

$$\mathbf{F}_i = 0, \quad i = 1, 2, \ldots N \tag{8.18}$$

These equations are not very useful in this form if the system is subjected to constraints, because the force $\mathbf{F}_i$ consists of the applied force and the force of constraints which is not known a priori, as we have seen in Section 8.2. We may calculate, however, the virtual work done on the system which, because of (8.18), is equal to zero:

$$\delta W = \sum_{i=1}^{N} \mathbf{F}_i \cdot \delta \mathbf{r}_i = 0$$

In terms of the generalized coordinates we obtain

$$\delta W = \sum_{j=1}^{f} Q_j \delta q_j = 0 \tag{8.19}$$

Because the q_j are all independent of each other, Eq. 8.19 can hold only if the generalized forces vanish separately. We conclude that a system of particles is in equilibrium if

$$Q_j = 0, \quad j = 1, 2, \ldots f \tag{8.20}$$

Example 8.9

Two masses m_1 and m_2 connected by a massless cord of length l are allowed to move without friction on two inclined planes under the influence of the gravitational force in the negative y-direction (see Figure 8.5). When is the system in equilibrium?

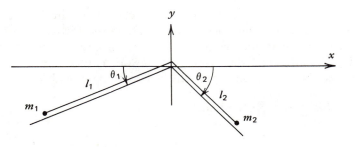

Figure 8.5 Example 8.9.

The number of degrees of freedom is one. We choose l_1 as the generalized coordinate. Because the forces $\mathbf{F}_1 = (0, -m_1 g)$ and $\mathbf{F}_2 = (0, -m_2 g)$ have no x-component, we need to calculate only

$$\frac{\partial y_1}{\partial l_1} = \frac{\partial}{\partial l_1}(-l_1 \sin \theta_1) = -\sin \theta_1$$

and

$$\frac{\partial y_2}{\partial l_1} = \frac{\partial}{\partial l_1}(-l_2 \sin \theta_2) = \frac{\partial}{\partial l_1}\left[-(l - l_1)\sin \theta_2\right] = \sin \theta_2$$

By Eq. 8.16 the generalized force associated with l_1 is

$$Q = \mathbf{F}_1 \cdot \frac{\partial \mathbf{r}_1}{\partial l_1} + \mathbf{F}_2 \cdot \frac{\partial \mathbf{r}_2}{\partial l_2} = F_{1y}\frac{\partial y_1}{\partial l_1} + F_{2y}\frac{\partial y_2}{\partial l_1}$$

$$= -m_1 g(-\sin \theta_1) - m_2 g \sin \theta_2,$$

and the equilibrium condition $Q = 0$ yields the familiar result

$$m_1 \sin \theta_1 = m_2 \sin \theta_2 \qquad \blacksquare$$

We assume now that the forces $\mathbf{F}_i$ are conservative. Then a potential energy function $V(\mathbf{r}_1, \ldots \mathbf{r}_N)$ exists such that

$$\mathbf{F}_i = -\nabla_i V,$$

see (7.26). By Eq. 8.9 the potential energy can be written as a function of the generalized coordinates, $V = V(q_1, \ldots q_f)$. It turns out that the generalized force Q_j is then the negative derivative of V with respect to the generalized coordinate q_j because

$$Q_j = \sum_{i=1}^{N} \mathbf{F}_i \cdot \frac{\partial \mathbf{r}_i}{\partial q_j} = -\sum_{i=1}^{N} \nabla_i V \cdot \frac{\partial \mathbf{r}_i}{\partial q_j} = -\frac{\partial V}{\partial q_j} \qquad (8.21)$$

8.5 KINETIC ENERGY

We need to take a closer look at the kinetic energy because it assumes a slightly more complicated form in Lagrange's formalism. As we shall see immediately, the kinetic energy T does not depend only on the generalized velocities; it is also, in general, a function of the generalized coordinates.

Using Eq. 8.11 we may write T in the form:

$$T = \sum_{i=1}^{N} \tfrac{1}{2} m_i \dot{\mathbf{r}}_i \cdot \dot{\mathbf{r}}_i$$

$$= \frac{1}{2} \sum_{i=1}^{N} m_i \left(\sum_{j=1}^{f} \frac{\partial \mathbf{r}_i}{\partial q_j} \dot{q}_j + \frac{\partial \mathbf{r}_i}{\partial t} \right) \cdot \left(\sum_{k=1}^{f} \frac{\partial \mathbf{r}_i}{\partial q_k} \dot{q}_k + \frac{\partial \mathbf{r}_i}{\partial t} \right)$$

$$= \frac{1}{2} \sum_{j=1}^{f} \sum_{k=1}^{f} \left(\sum_{i=1}^{N} m_i \frac{\partial \mathbf{r}_i}{\partial q_j} \cdot \frac{\partial \mathbf{r}_i}{\partial q_k} \right) \dot{q}_j \dot{q}_k$$

$$+ 2 \sum_{j=1}^{f} \left(\sum_{i=1}^{N} \frac{1}{2} m_i \frac{\partial \mathbf{r}_i}{\partial q_j} \cdot \frac{\partial \mathbf{r}_i}{\partial t} \right) \dot{q}_j + \sum_{i=1}^{N} \frac{1}{2} m_i \frac{\partial \mathbf{r}_i}{\partial t} \cdot \frac{\partial \mathbf{r}_i}{\partial t},$$

or

$$T = \frac{1}{2} \sum_{j=1}^{f} \sum_{k=1}^{f} a_{jk} \dot{q}_j \dot{q}_k + \sum_{j=1}^{f} b_j \dot{q}_j + c \tag{8.22}$$

where we have set

$$a_{jk} = \sum_{i=1}^{N} m_i \frac{\partial \mathbf{r}_i}{\partial q_j} \cdot \frac{\partial \mathbf{r}_i}{\partial q_k} \tag{8.23}$$

$$b_j = \sum_{i=1}^{N} m_i \frac{\partial \mathbf{r}_i}{\partial q_j} \cdot \frac{\partial \mathbf{r}_i}{\partial t} \tag{8.24}$$

$$c = \sum_{i=1}^{N} \frac{1}{2} m_i \frac{\partial \mathbf{r}_i}{\partial t} \cdot \frac{\partial \mathbf{r}_i}{\partial t} \tag{8.25}$$

The coefficients b_j and c are different from zero when rheonomous constraints are present, because in this case $\partial \mathbf{r}_i / \partial t \neq 0$ for at least one vector $\mathbf{r}_i$. In the absence of rheonomous constraints the kinetic energy is a homogeneous quadratic form in the generalized velocities $\dot{q}_j$:

$$T = \frac{1}{2} \sum_{j=1}^{f} \sum_{k=1}^{f} a_{jk} \dot{q}_j \dot{q}_k \tag{8.26}$$

The coefficients a_{ij} form a symmetric matrix that can be seen from the definition (8.23). The a_{ij} may be functions of the generalized coordinates.

Example 8.10

In plane polar coordinates, $q_1 = r$, $q_2 = \varphi$, the kinetic energy is $T = (1/2)m(\dot{r}^2 + r^2\dot{\varphi}^2)$. Here we have $a_{11} = m$, $a_{22} = mr^2$, and $a_{12} = a_{21} = 0$. ∎

As was the case for the generalized forces, it is often much simpler to derive the kinetic energy from the physical meaning of the generalized coordinates rather than to calculate T from Eqs. 8.22 through 8.25 directly. Another simplification occurs when, for a single mass point, orthogonal coordinates are chosen. The off-diagonal elements of a_{ij} vanish, and the kinetic energy becomes simply

$$T = \frac{1}{2} \sum_{j=1}^{f} a_{jj}\dot{q}_j^2$$

(see Example 8.10).

If T is a homogeneous quadratic form in the generalized velocities, it follows from (8.26) that

$$\sum_{l=1}^{f} \dot{q}_l \frac{\partial T}{\partial \dot{q}_l} = \frac{1}{2} \sum_{l=1}^{f} \sum_{j=1}^{f} \sum_{k=1}^{f} \dot{q}_l a_{jk}\left(\frac{\partial \dot{q}_j}{\partial \dot{q}_l}\dot{q}_k + \dot{q}_j \frac{\partial \dot{q}_k}{\partial \dot{q}_l}\right)$$

$$= \frac{1}{2} \sum_{j=1}^{f} \sum_{k=1}^{f} a_{jk} \sum_{l=1}^{f} \dot{q}_l(\delta_{jl}\dot{q}_k + \dot{q}_j\delta_{ke})$$

$$= \frac{1}{2} \sum_{j=1}^{f} \sum_{k=1}^{f} a_{jk}(\dot{q}_j\dot{q}_k + \dot{q}_j\dot{q}_k)$$

$$= 2T \tag{8.27}$$

This is a special case of a theorem by Euler, which states that for any homogeneous form f of degree n in the variables x_i the identity

$$\sum_i x_i \frac{\partial f}{\partial x_i} = nf$$

holds. Equation 8.27 will be needed later.

8.6 LAGRANGE'S EQUATIONS

We are now prepared to pursue one of the main objectives of this chapter, namely, to derive Lagrange's equations.

Again, we consider a system of N particles for which f generalized coordinates have been chosen. The system may or may not be subjected to constraints, which we assume to be holonomic. We begin with Newton's law of motion

$$m_i \frac{d}{dt}\left(\frac{d\mathbf{r}_i}{dt}\right) = \mathbf{F}_i, \quad i = 1,\dots N \tag{8.28}$$

Taking the inner produce of this vector equation with $\partial \mathbf{r}_i / \partial q_j$, and extending the sum over all particles we obtain

$$\sum_{i=1}^{N} m_i \frac{d}{dt}\left(\frac{d\mathbf{r}_i}{dt}\right) \cdot \frac{\partial \mathbf{r}_i}{\partial q_j} = \sum_{i=1}^{N} \mathbf{F}_i \cdot \frac{\partial \mathbf{r}_i}{\partial q_j}, \quad j = 1, \ldots f \tag{8.29}$$

The term on the right side is by (8.16) just the generalized force Q_j. In order to change the left-hand side (LHS) of (8.29) into a simpler expression, we first make use of the identity

$$\dot{u}v = (uv)^{\cdot} - u\dot{v}$$

of differential calculus. We find

$$\text{LHS} = \sum_{i=1}^{N} m_i \left[\frac{d}{dt}\left(\frac{d\mathbf{r}_i}{dt} \cdot \frac{\partial \mathbf{r}_i}{\partial q_j}\right) - \frac{d\mathbf{r}_i}{dt} \cdot \frac{d}{dt}\left(\frac{\partial \mathbf{r}_i}{\partial q_j}\right) \right]$$

Next, applying the identity (8.12) to the first term and the identity (8.13) to the second term, we obtain

$$\text{LHS} = \sum_{i=1}^{N} m_i \left[\frac{d}{dt}\left(\dot{\mathbf{r}}_i \cdot \frac{\partial \dot{\mathbf{r}}_i}{\partial \dot{q}_j}\right) - \dot{\mathbf{r}}_i \cdot \frac{\partial \dot{\mathbf{r}}_i}{\partial q_j} \right]$$

which is equal to

$$\text{LHS} = \sum_{i=1}^{N} m_i \left[\frac{d}{dt}\frac{1}{2}\frac{\partial}{\partial \dot{q}_j}(\dot{\mathbf{r}}_i \cdot \dot{\mathbf{r}}_i) - \frac{1}{2}\frac{\partial}{\partial q_j}(\dot{\mathbf{r}}_i \cdot \dot{\mathbf{r}}_i) \right]$$

$$= \frac{d}{dt}\frac{\partial}{\partial \dot{q}_j}\left(\sum_{i=1}^{N} \frac{1}{2}m_i\dot{\mathbf{r}}_i \cdot \dot{\mathbf{r}}_i\right) - \frac{\partial}{\partial q_j}\left(\sum_{i=1}^{N} \frac{1}{2}m_i\dot{\mathbf{r}}_i \cdot \dot{\mathbf{r}}_i\right)$$

$$= \frac{d}{dt}\frac{\partial}{\partial \dot{q}_j}T - \frac{\partial}{\partial q_j}T$$

where $T = (1/2)\sum_{i=1}^{N} m_i\dot{\mathbf{r}}_i \cdot \dot{\mathbf{r}}_i$ is the total kinetic energy of the system. The set of equations (8.29) then becomes

$$\frac{d}{dt}\frac{\partial}{\partial \dot{q}_j}T - \frac{\partial}{\partial q_j}T = Q_j, \quad j = 1, 2, \ldots f \tag{8.30}$$

These are Lagrange's equations. They consist of f second-order differential equations in the f variables q_j. The initial values of q_j and $\dot{q}_j$, or equivalent information, must be given to determine the solution uniquely.

Because the derivation of Lagrange's equations (8.30) from Newton's equations (8.28) did not involve any assumptions concerning the nature of the forces, Lagrange's equations, like Newton's, are applicable both to conservative and nonconservative systems. If the system of particles is not subjected to any constraints, Eqs. 8.30 are completely equivalent to Eqs. 8.28. If the system of particles is subjected to holonomic constraints (whether they are rheonomous or not), Lagrange's equations differ from Newton's in their dynamical content only insofar as the forces of constraints no longer appear. As a consequence of the absence of the forces of constraints, Lagrange's equations are equations of motion of the system in terms of the f independent generalized coordinates. There is, however, a formal difference between Eqs. 8.28 and 8.30. Although Eqs. 8.28 are vector equations, the concept of vector is hidden in Eqs. 8.30.

Another point needs to be made. Lagrange's equations assume the same *form* for any choice of the generalized coordinates. This does not mean that the equations themselves are the same for any choice of the generalized coordinates. For example, the equations of the motion in terms of Cartesian coordinates are obviously different from those expressed in terms of polar coordinates. The above statement means that the dynamical content of the equations is not altered if other generalized coordinates are selected or, equivalently, if a given set $\{q_1, \ldots q_f\}$ of generalized coordinates is transformed into another set $\{q_1', \ldots q_f'\}$. Equations that do not change the form if a coordinate transformation is performed are said to be form invariant. The concept of form invariance plays an important role in relativity.

If the system is conservative, we may replace the generalized force Q_j by the derivative $-\partial V/\partial q_j$ of the potential energy V.

Equations 8.30 can be written

$$\frac{d}{dt}\frac{\partial}{\partial \dot{q}_j}T - \frac{\partial}{\partial q_j}T + \frac{\partial}{\partial q_j}V = 0 \tag{8.31}$$

Because V does not depend on the (generalized) velocities, we can add

$$0 = -\frac{d}{dt}\frac{\partial V}{\partial \dot{q}_j}$$

to the left side of (8.31), which then becomes

$$\frac{d}{dt}\frac{\partial}{\partial \dot{q}_j}(T - V) - \frac{\partial}{\partial q_j}(T - V) = 0$$

Introducing the Lagrangian (or Lagrange function) L by

$$L = L(q_j, \dot{q}_j, t) = T - V \tag{8.32}$$

we obtain

$$\frac{d}{dt}\frac{\partial}{\partial \dot{q}_j}L - \frac{\partial}{\partial q_j}L = 0, \quad j = 1, 2, \ldots f \tag{8.33}$$

Sometimes one defines the Lagrange operator

$$\Lambda_j = \frac{d}{dt}\frac{\partial}{\partial \dot{q}_j} - \frac{\partial}{\partial q_j} \tag{8.34}$$

Equation 8.33 then assumes the simple form

$$\Lambda_j L = 0, \quad j = 1, 2, \ldots f \tag{8.35}$$

The derivation of Lagrange's equations (8.33) was not simple, but we are highly rewarded in many ways. First of all, the surprising feature of (8.33) is that all the dynamical information about a conservative system is contained in one scalar function, the Lagrangian L. Second, Lagrange's formalism provides us with a quick and elegant method to set up the equations of motion. We simply follow the three steps:

Step 1. Select suitable generalized coordinates.
Step 2. Find the Lagrangian in terms of the chosen generalized coordinates and velocities.
Step 3. Set up Eqs. 8.33.

Third, as will be shown in the next section, some conserved dynamical quantities may be derived from the Lagrangian in a simple way.

Let us now discuss three examples to familiarize ourselves with the Lagrangian method.

Example 8.11

Set up Lagrange's equations for a free particle. We choose the Cartesian coordinates x, y, z as the generalized coordinates. Because the particle is free, we have $V = 0$ and the Lagrangian becomes

$$L = T = \tfrac{1}{2} m \left(\dot{x}^2 + \dot{y}^2 + \dot{z}^2 \right)$$

Because

$$\frac{\partial L}{\partial \dot{x}} = m \dot{x}, \quad \frac{\partial L}{\partial x} = 0$$

and equivalent expressions for y and z, Lagrange's equations are

$$\Lambda_x L = \frac{d}{dt} \frac{\partial L}{\partial \dot{x}} - \frac{\partial L}{\partial x} = \frac{d}{dt} m \dot{x} - 0 = m \ddot{x} = 0,$$

$$\Lambda_y L = m \ddot{y} = 0 \quad \text{and} \quad \Lambda_z L = m \ddot{z} = 0 \qquad \blacksquare$$

Example 8.12

Set up Lagrange's equations for the system described in Example 8.9. We again choose l_1 as the generalized coordinate. We easily find

$$V = V_1 + V_2 = -m_1 g l_1 \sin \theta_1 - m_2 g (l - l_1) \sin \theta_2$$

and

$$T = T_1 + T_2 = \tfrac{1}{2}m_1\dot{l}_1^2 + \tfrac{1}{2}m_2[(l - l_1)\,{}^\bullet]^2 = \tfrac{1}{2}(m_1 + m_2)\dot{l}_1^2$$

such that

$$L = T - V = \tfrac{1}{2}(m_1 + m_2)\dot{l}_1^2 + g[m_1 l_1 \sin\theta_1 + m_2(l - l_1)\sin\theta_2]$$

The derivatives of L are

$$\frac{\partial L}{\partial \dot{l}_1} = (m_1 + m_2)\dot{l}_1 \quad \text{and} \quad \frac{\partial L}{\partial l_1} = g(m_1 \sin\theta_1 - m_2 \sin\theta_2),$$

and Lagrange's equation for l_1 reads

$$\Lambda L = (m_1 + m_2)\ddot{l}_1 - g(m_1 \sin\theta_1 - m_2 \sin\theta_2) = 0 \qquad \blacksquare$$

Example 8.13

Two mass points m_1 and m_2 are connected by a cord of length l which passes through a hole in a horizontal table. Mass m_1 moves without friction on the table, and mass m_2 acts as the bob of a simple pendulum (see Figure 8.6). Set up Lagrange's equations.

There are three holonomic conditions: (1) m_1 is confined to move in a horizontal plane, (2) m_2 is confined to move in a vertical plane, and (3) $r + s = l = $ constant. The number of degrees of freedom is $3N - c = 3 \times 2 - 3 = 3$. We choose the variables r, θ, and φ as the generalized coordinates. The following statements are self-explanatory:

$$V = V_1 + V_2 = 0 - m_2 gr \cos\theta = -m_2 gr \cos\theta$$

$$T = T_1 + T_2 = \tfrac{1}{2}m_1(\dot{r}^2 + r^2\dot{\theta}^2) + \tfrac{1}{2}m_2(\dot{s}^2 + s^2\dot{\varphi}^2)$$

$$= \tfrac{1}{2}m_1(\dot{r}^2 + r^2\dot{\theta}^2) + \tfrac{1}{2}m_2(\dot{r}^2 + (l - r)^2\dot{\varphi}^2)$$

$$L = T - V = \tfrac{1}{2}m_1(\dot{r}^2 + r^2\dot{\theta}^2) + \tfrac{1}{2}m_2(\dot{r}^2 + (l - r)^2\dot{\varphi}^2) + m_2 gr \cos\theta$$

$$\frac{\partial L}{\partial \dot{r}} = (m_1 + m_2)\dot{r}, \quad \frac{\partial L}{\partial r} = m_1 r\dot{\theta}^2 - m_2(l - r)\dot{\varphi}^2 + m_2 g \cos\theta$$

$$\frac{\partial L}{\partial \dot{\theta}} = m_1 r^2 \dot{\theta}, \quad \frac{\partial L}{\partial \theta} = -m_2 gr \sin\theta$$

$$\frac{\partial L}{\partial \dot{\varphi}} = m_2(l - r)^2 \dot{\varphi}, \quad \frac{\partial L}{\partial \varphi} = 0$$

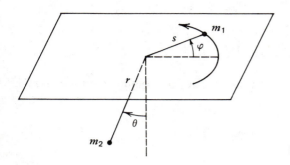

Figure 8.6 Example 8.13.

Lagrange's equations are

$$\Lambda_r L = (m_1 + m_2)\ddot{r} - m_1 r\dot{\theta}^2 + m_2(l - r)\dot{\varphi}^2 - m_2 g\cos\theta = 0$$

$$\Lambda_\theta L = \frac{d}{dt}\left[m_1 r^2\dot{\theta}\right] + m_1 gr\sin\theta = 0$$

$$\Lambda_\varphi L = \frac{d}{dt}\left[m_2(l - r)^2\dot{\varphi}\right] = 0$$

(Remember that the differential operator d/dt acts on all variables to the right of it and not only on the velocities. Thus, $(d/dt)(r^2\dot{\theta})$ is equal to $2r\dot{r}\dot{\theta} + r^2\ddot{\theta}$ and not merely $r^2\ddot{\theta}$.) ∎

If, in addition to conservative forces, nonconservative forces $\mathbf{F}_i'$, $i = 1, 2, \ldots N$, act on the system, Lagrange's equations become

$$\Lambda_j L = Q_j',$$

where

$$Q_j' = \sum_{i=1}^{N} \mathbf{F}_i' \cdot \frac{\partial \mathbf{r}_i}{\partial q_j}$$

is the generalized nonconservative force associated with q_j.

8.7 GENERALIZED MOMENTUM AND IGNORABLE COORDINATES

If the kinetic energy of a single mass point is written in Cartesian coordinates, the derivatives of $T = (1/2)m(\dot{x}^2 + \dot{y}^2 + \dot{z}^2)$ with respect to the velocities are just the coordinates of the momentum vector $\mathbf{p}$:

$$\frac{\partial T}{\partial \dot{x}} = m\dot{x} = p_x, \quad \frac{\partial T}{\partial \dot{y}} = m\dot{y} = p_y, \quad \frac{\partial T}{\partial \dot{z}} = m\dot{z} = p_z$$

Because the potential energy does not depend on the velocity, we also have

$$\frac{\partial L}{\partial \dot{x}} = p_x, \quad \frac{\partial L}{\partial \dot{y}} = p_y, \quad \frac{\partial L}{\partial \dot{z}} = p_z$$

This result suggests that we introduce the generalized (or conjugate) momentum p_j associated with q_j by

$$p_j = \frac{\partial L}{\partial \dot{q}_j} \tag{8.36}$$

One often refers to the variables (q_j, p_j) as the pair of conjugate (or canonical) coordinates and momenta. Because q_j does not necessarily have the dimension of a length, p_j does not necessarily have the dimension of linear momentum.

Example 8.14

In two dimensions, what are the generalized momenta associated with the plane polar coordinates r and φ?

We have $L = T - V = (1/2)m(\dot{r}^2 + r^2\dot{\varphi}^2) - V(r, \varphi)$. Then

$$p_r = \frac{\partial L}{\partial \dot{r}} = m\dot{r}$$

is the radial momentum of the particle, and

$$p_\varphi = \frac{\partial L}{\partial \dot{\varphi}} = mr^2\dot{\varphi}$$

is the angular momentum of the particle. ∎

Using Eq. 8.36 we may write Lagrange's equations (8.33) in the form

$$\frac{d}{dt}p_j = \frac{\partial L}{\partial q_j} \tag{8.37}$$

which bears great resemblance to Newton's equations (8.28). The left side is the rate of change of the generalized momentum. The right side represents the force that consists of two parts

$$\frac{\partial L}{\partial q_j} = \frac{\partial T}{\partial q_j} - \frac{\partial V}{\partial q_j}$$

We recognize the term $-\partial V/\partial q_j$ as the generalized conservative force. The other term $\partial T/\partial q_j$ arises from the motion of the particle (or systems of particles) itself, and it is sometimes called the fictitious force associated with q_j. [Usually, the forces arising from accelerated (noninertial) frames are called fictitious forces. But these comprise a larger class of forces and are either of the form $\partial T/\partial q_j$ or are proportional to second-order time derivatives of the coordinates.]

Example 8.15

Do fictitious forces associated with plane polar coordinates exist?

The kinetic energy is $T = (1/2)m(\dot{r}^2 + r^2\dot{\varphi}^2)$. The derivative $\partial T/\partial r = mr\dot{\varphi}^2(\geqslant 0)$ is the centrifugal force. It points away from the origin of the coordinate system. There is no fictitious force associated with φ because $\partial T/\partial\varphi = 0$. ∎

There are many instances in which the Lagrangian L does not depend on one or several of the chosen generalized coordinates while it is still a function of the associated generalized velocities. Coordinates of this kind are called cyclic or ignorable. Thus, if q_m is an ignorable coordinate, we have

$$\frac{\partial L}{\partial q_m} \equiv 0,$$

and it follows from (8.37) that $\dot{p}_m = 0$ or

$$p_m = \text{constant}$$

The generalized momentum associated with an ignorable coordinate is conserved.

Example 8.16

The coordinate φ in Example 8.11 is ignorable and the associated (angular) momentum is conserved:

$$m_2(l - r)^2\dot{\varphi} = \text{constant} \qquad ∎$$

The presence of ignorable coordinates is a great advantage because one integration of Lagrange's equations can be performed immediately for every ignorable coordinate. The difficulty of solving Lagrange's equations could, therefore, be greatly reduced if all the generalized coordinates were ignorable. The same statement applies to Hamilton's equations because, as we shall see in the next section, the property of being cyclic persists in Hamilton's formalism. Unfortunately, the problem of making all the f generalized coordinates ignorable is equivalent to that of finding the *complete* solution of a partial differential equation, which has been derived by Hamilton and Jacobi, but only in special cases can it be treated without excessive difficulties. A discussion of the Hamilton-Jacobi theory goes beyond the scope of this book.

8.8 HAMILTON'S EQUATIONS

Thus far we have investigated two formalisms of mechanics, namely, those of Newton and Lagrange. A third formalism exists which has been devised by Hamilton. A knowledge of Hamilton's formalism is fundamental to the understanding of statistical mechanics, the Hamilton-Jacobi theory, geometrical optics, and quantum mechanics. We shall touch on only a few aspects of Hamilton's formalism because most of the material belongs to advanced mechanics. Before we derive Hamilton's equations we need to first introduce the concept of the Legendre transformation.[5]

Let f be a function of two variables, x and y. The differential df of $f = f(x, y)$ is

$$df = \frac{\partial f}{\partial x} dx + \frac{\partial f}{\partial y} dy = u \, dx + v \, dy$$

where we have set $u = \partial f / \partial x$ and $v = \partial f / \partial y$. Now we consider the function

$$g = f - ux \tag{8.38}$$

The differential

$$dg = df - d(ux)$$

$$= u \, dx + v \, dy - (du \, x + u \, dx)$$

$$= -x \, du + v \, dy$$

is a perfect differential if g is a function of u and y. It then follows that

$$\frac{\partial g}{\partial u} = -x \quad \text{and} \quad \frac{\partial g}{\partial y} = v$$

The transformation (8.38) of $f(x, y)$ into $g(\partial f / \partial x, y)$ is called a Legendre transformation; g is called the Legendre transform of f with respect to the variable x. Other Legendre transforms of f are

$$h(x, v) = f - vy \quad \text{such that} \quad dh = u \, dx - y \, dv$$

and

$$k(u, v) = f - ux - vy \quad \text{such that} \quad dk = -x \, du - y \, dv$$

Each Legendre transformation has an inverse. For example, the inverse of (8.38) is given by $f = g + ux$. The concept of the Legendre transformation can easily be extended to functions of more than two variables.

We now introduce the Hamiltonian (or Hamilton function) H as the (negative) Legendre transform of the Lagrangian $L(q_j, \dot{q}_j, t)$ with respect to all generalized velocities,

$$H\left(q_j, \frac{\partial L}{\partial \dot{q}_j}, t\right) = -L + \sum_{j=1}^{f} \frac{\partial L}{\partial \dot{q}_j} \dot{q}_j, \tag{8.39}$$

[5]Legendre transformations also play an important role in thermodynamics.

or, replacing $\partial L / \partial \dot{q}_j$ by p_j,

$$H(q_j, p_j, t) = -L + \sum_{j=1}^{f} p_j \dot{q}_j \tag{8.40}$$

H is a function of the generalized coordinates q_j, the generalized momenta $p_j = \partial L / \partial \dot{q}_j$, and the time t. From (8.40) we obtain the differential of H in the two forms

$$dH = \sum_{j=1}^{f} \frac{\partial H}{\partial q_j} dq_j + \sum_{j=1}^{f} \frac{\partial H}{\partial p_j} dp_j + \frac{\partial H}{\partial t} dt \tag{8.41}$$

and

$$dH = -\sum_{j=1}^{f} \frac{\partial L}{\partial q_j} dq_j - \sum_{j=1}^{f} \frac{\partial L}{\partial \dot{q}_j} d\dot{q}_j - \frac{\partial L}{\partial t} dt$$
$$+ \sum_{j=1}^{f} \dot{q}_j dp_j + \sum_{j=1}^{f} p_j d\dot{q}_j$$

The second and the last terms cancel each other by the definition of p_j. Applying Lagrange's equations in the form (8.37) to the first term, we find

$$dH = -\sum_{j=1}^{f} \dot{p}_j dq_j + \sum_{j=1}^{f} \dot{q}_j dp_j - \frac{\partial L}{\partial t} dt \tag{8.42}$$

Comparing the coefficients of the differentials of the independent variables in Eqs. 8.41 and 8.42 we obtain Hamilton's equations

$$\dot{q}_j = \frac{\partial H}{\partial p_j}, \qquad -\dot{p}_j = \frac{\partial H}{\partial q_j}, \qquad j = 1, 2, \ldots f \tag{8.43}$$

and the relation

$$\frac{\partial H}{\partial t} = -\frac{\partial L}{\partial t} \tag{8.44}$$

Example 8.17

What are Hamilton's equations for a mass point moving on a straight line?
From the Lagrangian $L = T - V = (1/2)m\dot{x}^2 - V(x)$ we obtain

$$p = \frac{\partial L}{\partial \dot{x}} = m\dot{x} \qquad \text{or} \qquad \dot{x} = \frac{p}{m}$$

The Hamiltonian is then

$$H = -L + \frac{\partial L}{\partial \dot{x}} \dot{x} = -\tfrac{1}{2}m\dot{x}^2 + V(x) + m\dot{x} \cdot \dot{x} = \tfrac{1}{2}m\dot{x}^2 + V(x)$$

or

$$H(x, p) = \frac{1}{2m}p^2 + V(x)$$

Hamilton's equations read

$$\dot{x} = \frac{\partial H}{\partial p} = \frac{p}{m} \quad \text{and} \quad -\dot{p} = \frac{\partial H}{\partial x} = \frac{\partial V}{\partial x}$$

If we eliminate p using $\ddot{x} = \dot{p}/m$, we obtain the familiar result

$$\ddot{x} = -\frac{1}{m}\frac{\partial V}{\partial x} \qquad \blacksquare$$

Equations 8.43 represent a system of $2f$ first-order differential equations for the f generalized coordinates q_j and the f generalized momenta p_j. The initial values $q_j(0)$ and $p_j(0)$ or equivalent conditions must be prescribed. Because the inverse of the Legendre transformation (8.39) exists, and because we used Lagrange's equations to derive (8.42) and hence Hamilton's equations, these equations are equivalent to Lagrange's. The distinction between these two systems of equations lies entirely in the choice of the coordinates, a fact that has far-reaching consequences.

The system of Lagrange's equations consists of f differential equations for the f coordinates q_j. Obviously, $\dot{q}_j$ is uniquely determined by $q_j(t)$, but only after the equations of motion have been solved. In principle, the velocity $\dot{q}_j$ may take on any value. In this sense we may consider q_j and $\dot{q}_j$ as independent variables. Similarly, the f coordinates q_j and the f generalized momenta p_j may be treated as independent variables, although f of $2f$ Hamilton's equations establish relations between the p_j and the q_j. The other f equations are then differential equations for the f coordinates q_j.

If a generalized coordinate q_m is ignorable, then this property is not changed by the Legendre transformation (8.39), and the associated momentum p_m is a constant of the motion also in Hamilton's formalism. Occasionally, there is confusion about the meaning of $\partial H/\partial p_m$ if p_m = constant. Again, p_m is a variable before Hamilton's equations are solved. Only after the initial conditions have been applied to the solutions of Hamilton's equations does p_m assume the definite constant value $p_m(0)$. In other words, our knowledge of what will occur at a later stage must not interfere with what we do first.

One of the most beautiful properties of Hamilton's equations (8.43) is their perfect symmetry in the variables p_j and q_j, aside from the minus sign. This symmetry has been the starting point of many theoretical investigations, especially on canonical transformations and their applications to satellite orbits and to charged particles in electromagnetic fields. We must refer the reader to the literature about these subjects.

Let us now explore some of the properties of the Hamiltonian. It is easy to show (see Problem 8.15), that

$$\frac{dH}{dt} = \frac{\partial H}{\partial t} \qquad (8.45)$$

If H does not explicitly depend on the time, we have

$$\frac{dH}{dt} = 0,$$

and the Hamiltonian is a constant of the motion:

$$H(q_j, p_j) = \text{constant} \qquad (8.46)$$

Further, we have seen in Section 8.5 that the kinetic energy is a homogeneous quadratic form in the generalized velocities if rheonomous constraints are absent. Then, because of Euler's theorem (8.27), and because of the velocity independence of V, we obtain

$$
\begin{aligned}
H &= -L + \sum_{j=1}^{f} \frac{\partial L}{\partial \dot{q}_j} \dot{q}_j \\
&= -(T - V) + \sum_{j=1}^{f} \frac{\partial (T - V)}{\partial \dot{q}_j} \dot{q}_j \\
&= -T + V + 2T \\
&= T + V, \qquad (8.47)
\end{aligned}
$$

which is the total mechanical energy E. Thus, if the system is conservative, and if the holonomic constraints are time independent, Equation 8.47 provides us with a shortcut for constructing the Hamiltonian. We simply express the total energy in terms of the generalized coordinates and momenta. In all other cases the general definition (8.39) or (8.40) *must* be used for finding H. It must be emphasized that time independence of H and time independence of the holonomic constraints are two entirely different concepts. In Table 8.1 we compare the Hamiltonian with the total energy under various conditions.

In the next section we shall present an example where $H \neq E$ and $H = \text{constant}$.

Table 8.1 Hamiltonian and Total Energy Under Various Conditions

	Scleronomous Conditions, and V Does Not Depend on the Velocities	Rheonomous Conditions
H does not depend on t explicitly	$H = E$ $H = \text{constant}$	$H \neq E$ $H = \text{constant}$
H depends on t explicitly	$H = E$ $H \neq \text{constant}$	$H \neq E$ $H \neq \text{constant}$

8.9 CENTRAL FORCES REVISITED

Because of the importance of central isotropic forces we devote a special section to these forces in Lagrange's and Hamilton's formalisms.

Because the potential energy of the mass point depends only on the distance r between the force center and the mass point, we conveniently choose the spherical polar coordinates r, θ, and φ as the generalized coordinates. The Lagrangian takes the form

$$L = T - V = \tfrac{1}{2}m\left[\dot{r}^2 + r^2(\dot{\theta}^2 + \sin^2\theta\dot{\varphi}^2)\right] - V(r) \tag{8.48}$$

We immediately see that the azimuth angle φ is a cyclic coordinate, and therefore the associated generalized momentum p_φ is a constant of the motion:

$$p_\varphi = \frac{\partial L}{\partial \dot{\varphi}} = mr^2\sin^2\theta\dot{\varphi} = \text{constant} \tag{8.49}$$

The polar angle θ is not cyclic.

It would now be an unnecessary complication to derive Lagrange's equations in terms of all three coordinates r, θ, and φ. Because of angular momentum conservation the particle moves in a plane, which we define to be the x-y plane. We then have $\theta = \pi/2$, $\dot{\theta} = 0$, and the Lagrangian (8.48) reduces to

$$L = \tfrac{1}{2}m(\dot{r}^2 + r^2\dot{\varphi}^2) - V(r) \tag{8.50}$$

φ is still a cyclic coordinate, and (8.49) becomes

$$p_\varphi = mr^2\dot{\varphi} = \text{constant} \tag{8.51}$$

Lagrange's equation for r is

$$\Lambda_r L = m\ddot{r} - mr\dot{\varphi}^2 + \frac{\partial V}{\partial r} = 0$$

or

$$\ddot{r} - \frac{p_\varphi^2}{m^2 r^3} + \frac{1}{m}\frac{\partial V}{\partial r} = 0 \tag{8.52}$$

where we have used (8.51) to eliminate $\dot{\varphi}$. Equation 8.52 is exactly the radial equation for central forces derived in Chapter 6.

In order to obtain the Hamiltonian, we apply the equations

$$p_r = \frac{\partial L}{\partial \dot{r}} = m\dot{r}, \quad p_\theta = \frac{\partial L}{\partial \dot{\theta}} = mr^2\dot{\theta}, \quad p_\varphi = mr^2\sin^2\theta\dot{\varphi}$$

to eliminate the velocities in T. Because the system is conservative, we can use the shortcut $H = T + V$ such that

$$H = \frac{1}{2m}\left[p_r^2 + \frac{1}{r^2}\left(p_\theta^2 + \frac{1}{\sin^2\theta}p_\varphi^2 \right) \right] + V(r) \qquad (8.53)$$

We see that φ is cyclic in Hamilton's formalism also. For the motion in the x-y plane the Hamiltonian becomes

$$H = \frac{1}{2m}\left[p_r^2 + \frac{1}{r^2}p_\varphi^2 \right] + V(r) \qquad (8.54)$$

Hamilton's equations are then

$$\dot{r} = \frac{\partial H}{\partial p_r} = \frac{p_r}{m} \qquad (8.55)$$

$$-\dot{p}_r = \frac{\partial H}{\partial r} = -\frac{1}{mr^3}p_\varphi^2 + \frac{\partial V}{\partial r} \qquad (8.56)$$

$$\dot{\varphi} = \frac{\partial H}{\partial p_\varphi} = \frac{p_\varphi}{mr^2} \qquad (8.57)$$

and

$$-\dot{p}_\varphi = \frac{\partial H}{\partial \varphi} = 0 \qquad (8.58)$$

Substituting the time derivative of the radial momentum of (8.55) into (8.56) we again obtain the familiar equation (8.52).

We consider once more the two- (or three-) dimensional isotropic harmonic oscillator. If we choose polar coordinates as the generalized coordinates, we enjoy the fact that φ is cyclic and p_φ is a constant, but we have to pay a small price: The radial equation (8.52) which takes the form

$$\ddot{r} - \frac{p_\varphi^2}{m^2 r^3} + \frac{k}{m}r = 0$$

is, because of the extra term $p_\varphi^2/m^2 r^3$, more complicated than the two identical equations

$$\ddot{x} + \frac{k}{m}x = 0 \qquad \text{and} \qquad \ddot{y} + \frac{k}{m}y = 0 \qquad (8.59)$$

in Cartesian coordinates. If we choose Cartesian coordinates, neither x nor y is ignorable, and the associated generalized momenta are not conserved, which appears to be of no advantage. But the dynamical symmetry of the force law, namely, separability, is clearly exhibited by Eqs. 8.59, which imply that the total energies for the motion of the particles in the x-direction and in the y-direction are independent constants. The same conclusions can be drawn in Hamilton's formalism.

We shall now present an example which shows that the Hamiltonian is not always equal to the total energy.

Example 8.18

A mass point is subjected to a central isotropic force. Compare the Hamiltonian H of the mass point relative to a reference frame S fixed in space with the Hamiltonian H' of the same mass point relative to a frame S' rotating about the force center at constant angular speed ω.

The coordinates of the mass point in S are r and φ. The coordinates in S' are given by the rheonomous conditions

$$r' = r$$

$$\varphi' = \varphi - \omega t$$

In S we have

$$L = T - V = \tfrac{1}{2}m\left(\dot{r}^2 + r^2\dot{\varphi}^2\right) - V(r)$$

and

$$H = T + V = \frac{1}{2m}\left(p_r^2 + \frac{1}{r^2}p_\varphi^2\right) + V(r)$$

$$= E = \text{constant}$$

In S' we obtain

$$L' = T' - V' = \tfrac{1}{2}m\left(\dot{r}'^2 + r'^2\dot{\varphi}'^2\right) - V(r')$$

$$= \tfrac{1}{2}m\left(\dot{r}^2 + r^2(\dot{\varphi} - \omega)^2\right) - V(r),$$

hence,

$$p_r' = \frac{\partial L'}{\partial \dot{r}'} = m\dot{r} = p_r$$

and

$$p_\varphi' = \frac{\partial L'}{\partial \dot{\varphi}'} = mr'^2\dot{\varphi}' = mr^2(\dot{\varphi} - \omega) = p_\varphi - mr^2\omega$$

The Hamiltonian becomes

$$H' = H'(r', p_r', p_\varphi') = -L + \frac{\partial L'}{\partial \dot{r}'}\dot{r}' + \frac{\partial L'}{\partial \dot{\varphi}'}\dot{\varphi}'$$

$$= \frac{1}{2m}\left(p_r'^2 + \frac{1}{r'^2}p_\varphi'^2\right) + V(r')$$

$$= \frac{1}{2m}\left(p_r^2 + \frac{1}{r^2}\left(p_\varphi - mr^2\omega\right)^2\right) + V(r)$$

We see that $H' \neq E$, but, because H' is time independent, it is still a constant of the motion. Note that φ' is an ignorable coordinate. ∎

8.10 VELOCITY DEPENDENT POTENTIAL ENERGY

We have seen in Chapter 4 that velocity dependent forces which act in a direction perpendicular to the velocity leave the kinetic energy unchanged. Therefore, these forces may also be called "conservative" in a certain sense. The obvious question that arises is whether a scalar function exists from which the velocity dependent forces can be derived by some general prescription, as conservative forces can be derived from a potential energy function by differentiation. The way this might be done is suggested by Eq. 8.30,

$$\Lambda_j T = Q_j$$

If Q_j is conservative (in the strict sense), then

$$Q_j = -\frac{\partial V}{\partial q_j} = \frac{d}{dt}\frac{\partial V}{\partial \dot{q}_j} - \frac{\partial V}{\partial q_j} = \Lambda_j V$$

But here we have not used the velocity dependent part $(d/dt)(\partial/\partial \dot{q}_j)$ of the Lagrange operator Λ_j. We can make full use of the Lagrange operator to define the generalized potential energy function $U(q_k, \dot{q}_k)$ by

$$Q_j(q_k, \dot{q}_k) = \Lambda_j U(q_k, \dot{q}_k) \tag{8.60}$$

for forces that depend on the velocities also, provided that U exists. (Remember that not all position dependent forces are conservative, and that the existence of V is not always guaranteed.) Lagrange's equations can then be written

$$\Lambda_j T = \Lambda_j U \tag{8.61}$$

or, if we redefine the Lagrangian by

$$L = T - U, \tag{8.62}$$

Eqs. 8.61 read as before

$$\Lambda_j L = 0 \tag{8.63}$$

The generalized momentum

$$p_j = \frac{\partial L}{\partial \dot{q}_j} = \frac{\partial T}{\partial \dot{q}_j} - \frac{\partial U}{\partial \dot{q}_j} \tag{8.64}$$

now consists of two parts. In addition to the familiar term $\partial T/\partial \dot{q}_j$ there is another contribution associated with the generalized potential energy. If q_m is an ignorable coordinate, then the associated generalized momentum (8.64) is a constant of the motion.

Let us take a closer look at the Hamiltonian

$$H = -L + \sum_{j=1}^{f} \frac{\partial L}{\partial \dot{q}_j} \dot{q}_j$$

$$= -T + \sum_{j=1}^{f} \frac{\partial T}{\partial \dot{q}_j} \dot{q}_j + U - \sum_{j=1}^{f} \frac{\partial U}{\partial \dot{q}_j} \dot{q}_j \qquad (8.65)$$

If T is a homogeneous quadratic form in the velocities, the first two terms add up to T. If, further, the velocity dependent part of U is linear in the velocities,

$$U(q_k, \dot{q}_k) = V(q_k) + \sum_{j=1}^{f} u_j(q_k)\dot{q}_j, \qquad (8.66)$$

then we have

$$\sum_{j=1}^{f} \frac{\partial U}{\partial \dot{q}_j} \dot{q}_j = \sum_{j=1}^{f} u_j(q_k)\dot{q}_j, \qquad (8.67)$$

and the Hamiltonian (8.65) reduces to $H = T + V$, which is the total energy.

An important example of a velocity dependent force is the Lorentz force

$$\mathbf{F} = q(\mathbf{E} + \mathbf{v} \times \mathbf{B}) \qquad (8.68)$$

which a particle of electric charge q experiences when it moves through an electric field $\mathbf{E}$ and a magnetic field $\mathbf{B}$ (magnetic induction). It is shown in electromagnetic theory that $\mathbf{E}$ and $\mathbf{B}$ can be derived from a scalar potential ϕ and a vector potential $\mathbf{A}$ by

$$\mathbf{E} = -\nabla\phi - \frac{\partial \mathbf{A}}{\partial t} \quad \text{and} \quad \mathbf{B} = \nabla \times \mathbf{A} \qquad (8.69)$$

The generalized potential energy function $U(\mathbf{r}, \dot{\mathbf{r}})$ which yields the force (8.68) is then given by

$$U = q(\phi - \dot{\mathbf{r}} \cdot \mathbf{A}), \qquad (8.70)$$

(see Problem 8.21). The generalized momentum (8.64) becomes in this case

$$\mathbf{p} = m\dot{\mathbf{r}} + q\mathbf{A} \qquad (8.71)$$

and, because U is linear in the velocities, the Hamiltonian is

$$H = \frac{1}{2m}(\mathbf{p} - q\mathbf{A}) \cdot (\mathbf{p} - q\mathbf{A}) + q\phi \qquad (8.72)$$

The definition of the generalized potential energy U seems to be somewhat formal. The sole purpose of introducing U was to incorporate the velocity dependent forces into Lagrange's and Hamilton's formalism in a consistent way. However, in the example of the electromagnetic force much more has been achieved: The potential U, given by the expression (8.70), not $V = q\phi$, has the correct invariance property in relativity, and the Hamiltonian (8.72) is precisely the one needed to describe electromagnetic interactions in quantum mechanics.

PROBLEMS

8.1 Find several nontrivial examples of systems subjected to constraints that are:

(a) holonomic and scleronomous
(b) holonomic and rheonomous
(c) nonholonomic and scleronomous
(d) nonholonomic and rheonomous

8.2 Derive Eq. 8.8 by calculating the force on the mass point.

8.3 Two objects of masses m_1 and m_2, respectively, are connected by a massless cord of length l. The objects are allowed to move without friction on the inclined planes under the influence of the constant gravitational force in the negative y-direction as drawn in Figure 8.7. Find the equilibrium positions of the particles.

8.4 Two masses m_1 and m_2 are connected by a cord of constant length l. The mass points are free to move on the lower branch of the parabola $y^2 = 4ax$ under the influence of the constant gravitational force in the negative y-direction, while the cord passes through the focus of the parabola. Find the coordinates of the mass points in the equilibrium position.

[Remember that a parabola is the set of all points in a plane equidistant from a fixed point (the focus) and a fixed line (the directrix) in the plane.]

8.5 Find the kinetic energy of a particle in elliptical coordinates u and v, implicitly defined by $x = a \cosh u \cos v$ and $y = a \sinh u \sin v$, where a is a constant.

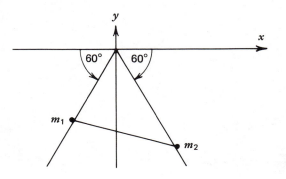

Figure 8.7 Problem 8.3.

8.6   Assume that the coordinate transformation between the Cartesian and the plane polar coordinates contains the time explicitly, for example, in the form $x = r\cos\varphi + at^2$, $y = r\sin\varphi + vt$, where a and v are constants. Find the kinetic energy in plane polar coordinates and verify that the kinetic energy is not homogeneous quadratic in the velocities $\dot{r}$ and $\dot{\varphi}$.

8.7   Find the kinetic energy of a particle in terms of the coordinates u and v defined by $u = \ln(r/a) - \varphi\cot b$ and $v = \ln(r/a) + \varphi\tan b$, where r and φ are plane polar coordinates and a and b are constants.

8.8   Derive Eq. 8.8 using the Lagrangian $L(x, \dot{x})$.

8.9   A chain of mass m and length l lies on a horizontal table with a portion hanging over the edge of the table. The only force acting on the system is the constant gravitational force in the negative y-direction. Set up and solve Lagrange's equation. Assume uniform mass density.

8.10 (a) Set up Lagrange's equations for the double pendulum (see Figure 8.8).
   (b) Assume that the amplitudes are very small. Simplify the equations keeping only terms linear in the variables.
   (c) Further simplify the equations for the two cases $m_1 \gg m_2$ and $m_1 \ll m_2$. Solve Lagrange's equations if possible, and describe the motion of the system.

8.11 An oscillating pendulum is a pendulum in which the cord has been replaced by a massless spring of spring constant k and relaxed length l_0.
   (a) Find the Lagrangian, and set up Lagrange's equations.
   (b) State the conditions under which Lagrange's equations reduce to the differential equations for the simple pendulum and for the linear harmonic oscillator.
   (c) Derive Hamilton's equations for the oscillating pendulum.

8.12 A point of mass m is sliding down a wedge of mass M which slides on a horizontal table without friction. Set up Lagrange's equations for the system.

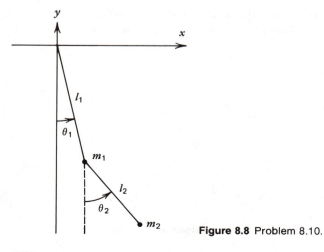

Figure 8.8 Problem 8.10.

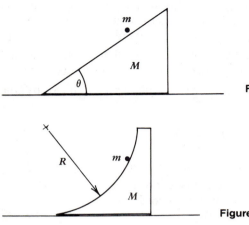

Figure 8.9a Problem 8.12a.

Figure 8.9b Problem 8.12b.

Consider the two cases:

(a) The surface of the wedge is an inclined plane (see Figure 8.9*a*).

(b) The surface of the wedge is a circle (see Figure 8.9*b*).

8.13 Two objects of masses m_1 and m_2, respectively, are connected to the ends of a massless cord of length *l*. The cord is free to move over two small frictionless pulleys separated by a distance *D*. The masses are subjected to the constant gravitational force. Consider the following motions:

(a) The masses move vertically (Atwood machine).

(b) Mass m_1 moves vertically, mass m_2 is the bob of a simple pendulum.

(c) Both masses move as simple pendulums.

(d) Both masses move as spherical pendulums.

(e) There is only one pulley. Mass m_1 moves as a simple pendulum and mass m_2 is free to move on a horizontal table without friction. (Set $D = 0$.)

For each system determine the number of degrees of freedom, choose the appropriate generalized coordinates, find the Lagrangian, derive Lagrange's equations, and identify the cyclic coordinates (if any).

8.14 Two masses m_1 and m_2 are fastened to the ends of a massless spring of spring constant *k* and relaxed length l_0. The system oscillates and rotates in a plane but moves freely through space otherwise.

(a) Find the number of degrees of freedom.

(b) Choose suitable generalized coordinates, find the Lagrangian, and derive Lagrange's equations. Note that this is a two-particle problem.

(c) Identify the cyclic coordinates, and state the corresponding conservation laws.

(d) Derive Hamilton's equations.

8.15 Derive the equation $dH/dt = \partial H/\partial t$.

8.16 Set up Hamilton's equations for the particle described in Example 8.18 in terms of the coordinates r' and φ'. Then write these equations in terms of r and φ.

8.17 Is the Legendre transformation of the Lagrangian with respect to the generalized *coordinates* q_j meaningful? What are the equations equivalent to Hamilton's? What happens if q_m is a cyclic coordinate?

8.18 Find Hamilton's equations for the anharmonic oscillator. The potential energy is $V = (1/2)kx^2 + (1/4)ax^4$, where k and a are constants.

8.19 Parabolic cylindrical coordinates (ξ, η, z) are implicitly defined by the equations

$$ x = \xi\eta, \quad y = \tfrac{1}{2}(\eta^2 - \xi^2), \quad z = z $$

Consider a three-dimensional isotropic harmonic oscillator.

(a) Derive Lagrange's equations in parabolic coordinates.

(b) Derive Hamilton's equations in parabolic coordinates.

8.20 Consider the Hamiltonian (8.65).

(a) When is H a constant of the motion?

(b) When is $H = T + V$?

8.21 (a) Derive Eqs. 8.68, 8.71, and 8.72 from the generalized potential energy $U = q(\phi - \mathbf{v} \cdot \mathbf{A})$.

(b) Set up Hamilton's equations.

chapter

9 Rigid Body Mechanics

Work is often the father of pleasure.

VOLTAIRE

9.1 INTRODUCTION

A rigid body is defined as a system of particles for which the distance between any pair of particles remains constant at all times. The definition of the rigid body can easily be extended to systems with a continuous mass distribution: The distance between any two infinitesimal volume elements and the mass in every volume element shall remain constant at all times. The concept of a rigid body is founded upon the two assumptions: that a force which acts on some point of the system does not cause any deformation of the body, and that the force is transmitted over the whole body instantaneously. These assumptions are reasonable approximations in many practical cases. But, all objects undergo deformations, however small, when they are subjected to stresses, and the idea of an infinite speed of transmission of a signal violates the principles of special relativity.

All macroscopic quantities that we introduced in Chapter 7 for systems of N interacting particles are also useful in rigid body mechanics:

Total mass:
$$M = \sum_{i=1}^{N} m_i \tag{9.1}$$

Center of mass vector:
$$\mathbf{R} = \frac{1}{M} \sum_{i=1}^{N} m_i \mathbf{r}_i \tag{9.2}$$

Total momentum:
$$\mathbf{P} = \sum_{i=1}^{N} m_i \dot{\mathbf{r}}_i \tag{9.3}$$

Total force:
$$\mathbf{F} = \mathbf{F}^{(e)} = \sum_{i=1}^{N} \mathbf{F}_i^{(e)} \tag{9.4}$$

Total angular momentum:
$$\mathbf{L} = \sum_{i=1}^{N} m_i \mathbf{r}_i \times \dot{\mathbf{r}}_i \tag{9.5}$$

Total torque:
$$\mathbf{N} = \mathbf{N}^{(e)} + \mathbf{N}^{(i)} = \sum_{i=1}^{N} \mathbf{r}_i \times \mathbf{F}_i^{(e)} + \sum_{i=1}^{N} \mathbf{r}_i \times \mathbf{F}_i^{(i)} \qquad (9.6)$$

Total kinetic energy:
$$T = \sum_{i=1}^{N} \frac{1}{2} m_i \dot{\mathbf{r}}_i \cdot \dot{\mathbf{r}}_i \qquad (9.7)$$

Also, the equations of motion for the rigid body are identical with Eqs. 7.16 and 7.35.

Because a nonvanishing internal torque has never been observed for a rigid body, it is reasonable to further assume that the internal forces $\mathbf{F}_{ij}$ satisfy Newton's third law in its strong form (7.13). The equations of motion can then be formulated solely in terms of the external forces and the external torques. Omitting the superscript "(e)" for "external" we have

$$\dot{\mathbf{P}} = \mathbf{F} \qquad (9.8)$$

and

$$\dot{\mathbf{L}} = \mathbf{N} \qquad (9.9)$$

We shall see that both $\mathbf{L}$ and T, and hence (9.9), can be written in a more convenient form in terms of the angular velocity vector and the moment of inertia tensor.

Clearly, the conservation laws

$$\Delta \mathbf{P} = 0 \qquad \text{if} \qquad \int_{t_1}^{t_2} \mathbf{F} \, dt = 0$$

and

$$\Delta \mathbf{L} = 0 \qquad \text{if} \qquad \int_{t_1}^{t_2} \mathbf{N} \, dt = 0$$

also apply to rigid bodies.

9.2 THE EQUATIONS OF MOTION

In order to describe the motion of a rigid body let us first introduce the following coordinate systems. The space system S is a coordinate system fixed in space (inertial frame). The body system S_B is a coordinate system fixed to the rigid body. Its origin O_B is attached to some point of the rigid body. The center of mass system S_{CM} is a special body system. Its origin is attached to the center of mass of the rigid body.

A simple counting procedure which involves the number of particles and the number of independent holonomic conditions on the particles reveals that the number of degrees of freedom of a rigid body is six (see Problem 9.1). We can choose as the six generalized coordinates the three coordinates of the point O_B of the rigid body relative to the space system, and the three angles of rotation about three different axes passing through O_B. It then follows that the general motion of a rigid body is a superposition of a pure translation and a pure rotation (Chasles's theorem). In this section we shall derive two independent sets of equations of motion, one for the pure translation and the other for the pure rotation. The existence of these sets of equations provides an explicit proof of Chasles's theorem. As a consequence of this theorem we expect that both L and T decompose each into an orbital and a rotational part.

The definition of the rigid body implies that all points move with the same instantaneous angular velocity about the instantaneous axis of rotation passing through O_B. But because vectors are quantities that do not depend on the chosen coordinate system, the angular velocity vector is independent of the reference point O_B (see also Problem 9.2). We can also state this result in the following form: Each point of the rigid body undergoes the same infinitesimal angular displacement that is independent of the reference point.

Let us now take a closer look at Eqs. 8.8 and 8.9. Equation 8.8 is the equation of the translational motion of the center of mass. This equation will not be discussed here because the general motion of a mass point has already been investigated in Chapter 4. Equation 8.9 is the equation of rotational motion of the rigid body in the space system because both L and N were defined by (8.5) and (8.6), respectively, relative to the origin of the space system. As we shall show now, however, an equation of the rotational motion can be extracted from (8.9), which is written in terms of L and N relative to the center of mass.

Let r_B be the radius vector of some point O_B of the rigid body in the space system, and let r_i' be the radius vector of a mass point P of the rigid body relative to O_B (see Figure 9.1). Clearly, the radius vector r_i of P relative to O is

$$r_i = r_B + r_i' \tag{9.10}$$

Because P rotates about O_B with the angular velocity ω, the velocity $\dot{r}_i$ of the ith mass point is

$$\dot{r}_i = \dot{r}_B + \omega \times r_i' \tag{9.11}$$

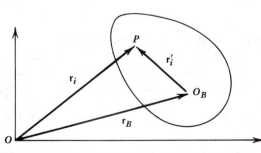

Figure 9.1 Arbitrarily chosen reference point on a rigid body.

Observe that the angular velocity $\boldsymbol{\omega}$ is measured in the fixed space system. Substituting the right sides of (9.10) and (9.11) for $\mathbf{r}_i$ and $\dot{\mathbf{r}}_i$ into (9.5) for the angular momentum we obtain

$$
\begin{aligned}
\mathbf{L} &= \sum_{i=1}^{N} m_i (\mathbf{r}_B + \mathbf{r}_i') \times (\dot{\mathbf{r}}_B + \dot{\mathbf{r}}_i') \\
&= \sum_{i=1}^{N} m_i \mathbf{r}_B \times \dot{\mathbf{r}}_B + \sum_{i=1}^{N} m_i \mathbf{r}_i' \times \dot{\mathbf{r}}_B \\
&\quad + \sum_{i=1}^{N} m_i \mathbf{r}_B \times (\boldsymbol{\omega} \times \mathbf{r}_i') + \sum_{i=1}^{N} m_i \mathbf{r}_i' \times (\boldsymbol{\omega} \times \mathbf{r}_i')
\end{aligned} \tag{9.12}
$$

$\mathbf{L}$ assumes the simplest form in the center of mass frame S_{CM}. Because

$$
\mathbf{r}_B = \mathbf{R}
$$

and because

$$
\sum_{i=1}^{N} m_i \mathbf{r}_i' = 0 \tag{9.13}
$$

in S_{CM} (see Eq. 7.41), the second and the third term of (9.12) vanish. Using Eqs. 9.1, 9.2, and 9.3, we can write $\mathbf{L}$ in the form

$$
\begin{aligned}
\mathbf{L} &= \mathbf{R} \times \mathbf{P} + \sum_{i=1}^{N} m_i \mathbf{r}_i' \times (\boldsymbol{\omega} \times \mathbf{r}_i') \\
&= \mathbf{R} \times \mathbf{P} + \mathbf{L}'
\end{aligned} \tag{9.14}
$$

The first term that is independent of $\boldsymbol{\omega}$ constitutes the orbital angular momentum of the rigid body. The second term

$$
\mathbf{L}' = \sum_{i=1}^{N} m_i \mathbf{r}_i' \times (\boldsymbol{\omega} \times \mathbf{r}_i') \tag{9.15}
$$

is the angular momentum of the rigid body relative to the center of mass (intrinsic or spin angular momentum). Note that the angular velocity $\boldsymbol{\omega}$ did not change under the translation of the origin of the coordinate system S_B from the arbitrary point O_B to the center of mass. The time derivative of the angular momentum is

$$
\dot{\mathbf{L}} = \mathbf{R} \times \dot{\mathbf{P}} + \dot{\mathbf{L}}' \tag{9.16}
$$

because

$$
\dot{\mathbf{R}} \times \mathbf{P} = 0
$$

Let us now rewrite the torque. Using Eqs. 9.10 and 9.13 we obtain

$$\mathbf{N} = \sum_{i=1}^{N} \mathbf{r}_i \times \mathbf{F}_i$$

$$= \sum_{i=1}^{N} \mathbf{R} \times \mathbf{F}_i + \sum_{i=1}^{N} \mathbf{r}_i' \times \mathbf{F}_i \qquad (9.17)$$

The first term is by definition (9.4) of the total force and by Eq. 9.8 equal to $\mathbf{R} \times \dot{\mathbf{P}}$. The second term

$$\mathbf{N}' = \sum_{i=1}^{N} \mathbf{r}_i' \times \mathbf{F}_i \qquad (9.18)$$

is the total torque relative to the center of mass. Then (9.17) becomes

$$\mathbf{N} = \mathbf{R} \times \dot{\mathbf{P}} + \mathbf{N}' \qquad (9.19)$$

Combining (9.9), (9.16), and (9.19) we obtain

$$\dot{\mathbf{L}}' = \mathbf{N}' \qquad (9.20)$$

which is the equation we sought.

Let us summarize: We have found two independent sets of dynamical equations for the rigid body. Equations 9.8 describe the translational motion of the center of mass relative to the space system S. They may be formulated in terms of the three (generalized) coordinates of the center of mass. Equations 9.20 describe the rotation of the rigid body about the center of mass. They may be formulated in terms of three independent angles of rotation of the rigid body relative to S. This completes the proof of Chasles's theorem.

9.3 EQUILIBRIUM OF A RIGID BODY

Equations 9.8 and 9.9 imply that a rigid body is in equilibrium if both the total force and the total torque relative to the origin of the space system vanish:

$$\mathbf{F} = \sum_{i=1}^{N} \mathbf{F}_i = 0, \qquad (9.21)$$

$$\mathbf{N} = \sum_{i=1}^{N} \mathbf{r}_i \times \mathbf{F}_i = 0 \qquad (9.22)$$

Substituting the right-hand side of (9.10) for $\mathbf{r}_i$ into (9.22) we obtain

$$\mathbf{N} = \sum_{i=1}^{N} \mathbf{r}_B \times \mathbf{F}_i + \sum_{i=1}^{N} \mathbf{r}_i' \times \mathbf{F}_i = 0$$

which because of (9.21) reduces to

$$\sum_{i=1}^{N} \mathbf{r}'_i \times \mathbf{F}_i = 0$$

Thus, if $\mathbf{F} = 0$ and if $\mathbf{N} = 0$ relative to some reference point in space, then there is also no net torque $\mathbf{N}'$ relative to any other reference point. In particular, if $\mathbf{r}_B = \mathbf{R}$, then the torque $\mathbf{N}'$ (Eq. 9.18), relative to the center of mass, vanishes. This special case also follows from (9.20) directly. For calculations of the stresses in static structures one usually chooses that reference point relative to which the torque assumes the simplest form.

The torque $\mathbf{N}'$ is zero if all forces $\mathbf{F}_i$ act in the same direction and are proportional to the masses m_i, because

$$\mathbf{N}' = \sum_{i=1}^{N} r'_i \times m_i \mathbf{a} = \sum_{i=1}^{N} m_i \mathbf{r}'_i \times \mathbf{a} = 0 \tag{9.23}$$

Thus, a rigid body falling freely in a uniform gravitational field, where $\mathbf{a} = \mathbf{g} =$ constant, does not experience a net torque.

That point of a rigid body, relative to which the torque produced by the gravitational forces vanishes, is called the center of gravity of the rigid body. The center of gravity coincides with the center of mass if the rigid body is placed in a uniform gravitational field.

9.4   THE MOMENT OF INERTIA

9.4.1   Definition of the Moment of Inertia
In this and the following sections the coordinate form of vectors, inner product of vectors, and tensors will be used.

Let us select an arbitrary point O_B of the rigid body as the origin of the body system S_B with Cartesian coordinates. At this stage we do not attempt to relate the coordinate axes to the geometrical symmetry properties of the body. The index i running from 1 to N shall refer to the particle, whereas the indices k, l, m, etc., running from 1 to 3 shall refer to the coordinate axes.

Henceforth, we shall use $\mathbf{r}_i$ instead of $\mathbf{r}'_i$ to denote the position vector of the ith mass point in the coordinate system S_B. The angular momentum of the rigid body about the point O_B can then be written

$$\mathbf{L} = \sum_{i=1}^{N} m_i \mathbf{r}_i \times (\boldsymbol{\omega} \times \mathbf{r}_i)$$

$$= \sum_{i=1}^{N} m_i \left(r_i^2 \boldsymbol{\omega} - (\mathbf{r}_i \cdot \boldsymbol{\omega}) \mathbf{r}_i \right)$$

where

$$r_i^2 = \mathbf{r}_i \cdot \mathbf{r}_i = x_i^2 + y_i^2 + z_i^2$$

The kth coordinate of **L** is given by

$$L_k = \sum_{i=1}^{N} m_i \left(r_i^2 \omega_k - \sum_{l=1}^{3} x_{i_k} x_{i_l} \omega_l \right) \tag{9.24}$$

The fact that the angular velocity is the same for all mass points of the rigid body allows us to factor the coordinates of the angular velocity such that L_k becomes

$$L_k = \sum_{l=1}^{3} \left\{ \sum_{i=1}^{N} m_i \left(r_i^2 \delta_{kl} - x_{i_k} x_{i_l} \right) \right\} \omega_l$$

$$= \sum_{l=1}^{3} I_{kl} \omega_l \tag{9.25}$$

The quantities

$$I_{kl} = \sum_{i=1}^{N} m_i \left(r_i^2 \delta_{kl} - x_{i_k} x_{i_l} \right) \tag{9.26}$$

are the components of a tensor I, called the tensor of the moment of inertia or shorter the inertia tensor of the rigid body relative to the coordinate system S_B. [Referring back to Section 2.10, we remind the reader that I is called a tensor because it satisfies certain conditions under a coordinate transformation (rotation). For our purpose it is not necessary to explore the transformation properties in detail.] Because the components of I form a 3×3 symmetric matrix, only six of the nine elements of I_{kl} are independent of each other. The diagonal elements

$$I_{kk} = \sum_{i=1}^{N} m_i \left(r_i^2 - x_{i_k}^2 \right) \tag{9.27}$$

are called the moments of inertia about the kth axis, the off-diagonal elements

$$I_{kl} = I_{lk} = - \sum_{i=1}^{N} m_i x_{i_k} x_{i_l} \qquad k \neq l \tag{9.28}$$

are called the products of inertia. In particular, for I_{11} we have

$$I_{11} = \sum_{i=1}^{N} m_i \left(x_{i_1}^2 + x_{i_2}^2 + x_{i_3}^2 - x_{i_1}^2 \right)$$

$$= \sum_{i=1}^{N} m_i \left(x_{i_2}^2 + x_{i_3}^2 \right) \tag{9.29}$$

The factor of m_i is the square of the distance from the x_1-axis to the ith mass point. Therefore, the diagonal elements I_{kk} are always positive or zero. I_{kk} is zero

Table 9.1 Moment of Inertia for Various Objects (Uniform Density)

Object	Dimensions	Volume	Axis of Rotation	Moment of Inertia	
Thin rod	a = length		Perpendicular to the rod		
			—through one end	$\frac{1}{3}Ma^2$	
			—through the CM	$\frac{1}{12}Ma^2$	
Thin circular ring	R = radius		Axis of cylindrical symmetry	MR^2	
			Through the CM in the ring plane	$\frac{1}{2}MR^2$	
Thin rectangular sheet	a = length b = width		Perpendicular to the sheet through the CM	$\frac{1}{12}M(a^2 + b^2)$	
Rectangular prism	sides a, b, c	abc	Axis of symmetry through the CM in the direction of side c	$\frac{1}{12}M(a^2 + b^2)$	
Circular cylinder	R = radius h = height	$\pi R^2 h$	Axis of cylindrical symmetry	$\frac{1}{2}MR^2$	
			Perpendicular to the cylindrical axis of symmetry through the CM	$\frac{1}{12}M(3R^2 + h^2)$	

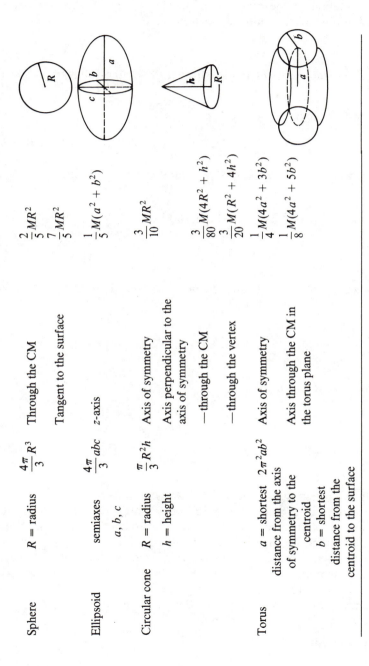

Sphere	R = radius	$\dfrac{4\pi}{3}R^3$	Through the CM	$\dfrac{2}{5}MR^2$
			Tangent to the surface	$\dfrac{7}{5}MR^2$
Ellipsoid	semiaxes a, b, c	$\dfrac{4\pi}{3}abc$	z-axis	$\dfrac{1}{5}M(a^2 + b^2)$
Circular cone	R = radius h = height	$\dfrac{\pi}{3}R^2h$	Axis of symmetry	$\dfrac{3}{10}MR^2$
			Axis perpendicular to the axis of symmetry	
			—through the CM	$\dfrac{3}{80}M(4R^2 + h^2)$
			—through the vertex	$\dfrac{3}{20}M(R^2 + 4h^2)$
Torus	a = shortest distance from the axis of symmetry to the centroid b = shortest distance from the centroid to the surface	$2\pi^2 ab^2$	Axis of symmetry	$\dfrac{1}{4}M(4a^2 + 3b^2)$
			Axis through the CM in the torus plane	$\dfrac{1}{8}M(4a^2 + 5b^2)$

if all mass points lie on the x_k-axis. The off-diagonal elements may be positive, zero, or negative. Moments of inertia for various objects are listed in Table 9.1. For the calculation of the moments of inertia of rigid bodies with continuous mass distributions, see the following section.

Example 9.1

What are the components of the inertia tensor for the following mass distribution?

$$m_i = 1 \quad \text{for} \quad i = 1, 2, 3, 4,$$

$$\mathbf{r}_1 = (1, 0, 0), \quad \mathbf{r}_2 = (1, 0, 1), \quad \mathbf{r}_3 = (-1, 0, 0), \quad \mathbf{r}_4 = (-1, 0, -1)$$

We have $r_1^2 = r_3^2 = 1$ and $r_2^2 = r_4^2 = 2$. Applying (9.27) we find

$$I_{11} = \sum_{i=1}^{4} m_i \left(r_i^2 - x_{i_1}^2 \right) = \sum_{i=1}^{4} r_i^2 - \sum_{i=1}^{4} x_{i_1}^2$$

$$= (1 + 2 + 1 + 2) - (1 + 1 + 1 + 1) = 2$$

Similarly, we obtain $I_{22} = 6$ and $I_{33} = 4$. Because $x_{i_2} = 0$ for all i, we have $I_{12} = I_{21} = I_{32} = I_{23} = 0$, whereas

$$I_{13} = I_{31} = -\sum_{i=1}^{4} m_i x_{i_1} x_{i_3} = -(1 \times 0 + 1 \times 1 + (-1) \times 0 + (-1)(-1)) = 2$$

The matrix I has the form

$$\mathsf{I} = \begin{pmatrix} 2 & 0 & -2 \\ 0 & 6 & 0 \\ -2 & 0 & 4 \end{pmatrix} \qquad \blacksquare$$

9.4.2 Rigid Bodies with Continuous Mass Distributions

Thus far we have considered only rigid bodies that consisted of N discrete mass points of masses m_i located at $\mathbf{r}_i$. The transition to continuous mass distributions requires only a slight change of the mathematical formalism. The transition does not affect the dynamical equations. All that needs to be done is to replace the sum over the masses m_i by integrals over the mass distribution dm. Thus, the expression for the total mass $M = \Sigma m_i$ goes over into $M = \int dm$.

It is more convenient to introduce the mass densities:

volume mass density: $\qquad\qquad \rho = \dfrac{dm}{dV}$ (9.30a)

surface mass density: $\qquad\qquad \sigma = \dfrac{dm}{da}$ (9.30b)

linear mass density:
$$\lambda = \frac{dm}{dl} \qquad (9.30c)$$

We then have

total mass:
$$M = \int_v \rho \, dV \qquad (9.31)$$

center of mass vector:
$$\mathbf{R} = \frac{1}{M} \int_v \mathbf{r}\rho \, dV \qquad (9.32)$$

moment of inertia:
$$I_{kl} = \int_v \left(r^2 \delta_{kl} - x_k x_l \right) \rho \, dV \qquad (9.33)$$

and similar integrals for surface and linear mass distributions. Although the volume integrals look simple the evaluation of these integrals may often prove to be very tedious.

Example 9.2

Find the moment of inertia of a rigid rod of uniform mass density, mass m and length l about an endpoint. We have $\lambda = m/l =$ constant such that

$$I = \int_0^l x^2 \lambda \, dx = \lambda \int_0^l x^2 \, dx = \frac{m}{l} \cdot \frac{1}{3} l^3 = \frac{1}{3} m l^2. \qquad \blacksquare$$

9.4.3   The Radius of Gyration

It is often convenient to introduce the radius of gyration k as that distance from the axis of rotation to a point where the mass M of the rigid body should be placed in order to produce the same moment of inertia. Thus, if the axis of rotation is the x_l-axis, we have $I_{ll} = Mk_l^2$, hence,

$$k_l = \sqrt{\frac{I_{ll}}{M}} \qquad (9.34)$$

Example 9.3

What are the radii of gyration for the system described in Example 9.1? Because $M = 4$, we obtain

$$k_1 = \sqrt{\frac{I_{11}}{M}} = \sqrt{\frac{2}{4}} = \sqrt{\frac{1}{2}}, \quad k_2 = \sqrt{\frac{I_{22}}{M}} = \sqrt{\frac{6}{4}} = \sqrt{\frac{3}{2}},$$

$$k_3 = \sqrt{\frac{I_{33}}{M}} = \sqrt{\frac{4}{4}} = 1 \qquad \blacksquare$$

9.4.4 The Parallel Axis Theorem

From now on it will be advantageous to write the inertia tensor in integral form. The formalism becomes simpler because particle indices no longer appear. The equations to be derived also hold for rigid bodies consisting of discrete particles.

We wish to derive a relation between the inertia tensor I, Eq. 9.33 (defined in the body system S_B), and the inertia tensor I′ with the components

$$I'_{kl} = \int_v \left(r'^2 \delta_{kl} - x'_k x'_l \right) \rho \, dV \tag{9.35}$$

(defined in the center of mass system S_{CM}). We assume that the Cartesian coordinate axes of S_B and S_{CM} are parallel to each other. Let $(x_{0_1}, x_{0_2}, x_{0_3})$ be the coordinates of the center of mass in S_B such that

$$x_k = x_{0_k} + x'_k, \quad k = 1, 2, 3 \tag{9.36}$$

The components of I can then be written

$$I_{kl} = \int_v \left[(\mathbf{r}_0 + \mathbf{r}') \cdot (\mathbf{r}_0 + \mathbf{r}') \delta_{kl} - \left(x_{0_k} + x'_k \right)\left(x_{0_l} + x'_l \right) \right] \rho \, dV$$

$$= \int_v \left(r_0^2 \delta_{kl} - x_{0_k} x_{0_l} \right) \rho \, dV + \int_v \left(r'^2 \delta_{kl} - x'_k x'_l \right) \rho \, dV$$

$$+ 2 \int_v \mathbf{r}_0 \cdot \mathbf{r}' \delta_{kl} \rho \, dV - x_{0_k} \int_v x'_l \rho \, dV - \int_v x'_k \rho \, dV \cdot x_{0_l}$$

Because the last three integrals vanish in S_{CM}, we obtain

$$I_{kl} = M \left(r_0^2 \delta_{kl} - x_{0_k} x_{0_l} \right) + I'_{kl} \tag{9.37}$$

Inspecting the diagonal elements we deduce that

$$I_{kk} = M \left(r_0^2 - x_{0_k}^2 \right) + I'_{kk}$$

$$= M d_k^2 + I'_{kk} \tag{9.38}$$

where $d_k^2 = r_0^2 - x_{0_k}^2$ is the square of the shortest distance from the axis of rotation to the center of mass. Equation 9.38 is known as the parallel axis theorem, also called Steiner's theorem. It states that the moment of inertia of a rigid body in some coordinate system about a given axis is equal to the moment of inertia of a mass point of mass M located at the center of mass about that axis, plus the moment of inertia of the body in the center of mass frame about an axis parallel to the first. Because $M d_k^2 \geq 0$, it follows that the moments of inertia I_{kk} assume minimum values in the center of mass frame.

Equation 9.38 is of great practical value because it often enables us to calculate the moment of inertia of an object more easily than by direct application of Eqs. 9.26 or 9.33.

Example 9.4

What is the moment of inertia of a sphere about an axis passing through a point at a distance of one-half the radius from the center, perpendicular to the line from the center to that point? From Table 9.1 we find $I' = (2/5)MR^2$. Because $d_k = R/2$, we obtain

$$I = M\left(\tfrac{1}{2}R\right)^2 + \tfrac{2}{5}MR^2 = \tfrac{13}{20}MR^2 \qquad \blacksquare$$

9.4.5 The Perpendicular Axis Theorem

A rigid body whose mass is distributed in a single plane is called a plane lamina. Suppose that this plane lamina lies in the x_1-x_2 plane such that $x_3 = 0$. The diagonal elements of I for a plane lamina are then

$$I_{11} = \sum_{i=1}^{N} m_i x_{i_2}^2 \qquad \text{or} \qquad I_{11} = \int x_2^2 \sigma \, da,$$

$$I_{22} = \sum_{i=1}^{N} m_i x_{i_1}^2 \qquad \text{or} \qquad I_{22} = \int x_1^2 \sigma \, da,$$

and

$$I_{33} = \sum_{i=1}^{N} m_i\left(x_{i_1}^2 + x_{i_2}^2\right) \qquad \text{or} \qquad I_{33} = \int \left(x_1^2 + x_2^2\right)\sigma \, da,$$

such that

$$I_{33} = I_{11} + I_{22} \qquad (9.39)$$

This equation is known as the perpendicular axis theorem. This theorem is also very useful for calculating the moment of inertia.

Example 9.5

Consider a square lamina of side a in the x_1-x_2 plane with the coordinate axes passing through the center of mass. The orientation of the square in the x_1-x_2 plane is arbitrary (see Figure 9.2). What is the moment of inertia about the x_1-axis? Because of symmetry we conclude that $I_{11} = I_{22}$. Table 9.1 gives us the value $I_{33} = (1/6)Ma^2$. Therefore, $I_{11} = (1/2)I_{33} = (1/12)Ma^2$. The result is independent of the orientation of the square in the x_1-x_2 plane. $\qquad \blacksquare$

9.4.6 Principal Moments of Inertia and Principal Axes

We mentioned previously that the inertia tensor I, Eq. 9.33, can be viewed as a symmetric matrix, which, in general, has six independent components in a given Cartesian coordinate system S_B. It follows then from the theorems stated in

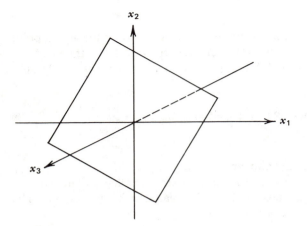

Figure 9.2 Example 9.5.

Section 2.12 that the inertia matrix has three real eigenvalues, and that the eigenvectors associated with these eigenvalues are orthogonal (or can be chosen orthogonal in the case of degeneracy). The eigenvalues of I, denoted by I_1, I_2, and I_3, are called the principal moments of inertia. The I_k are always positive or zero. We shall label the principal moments of inertia in such a way that

$$I_1 \leqslant I_2 \leqslant I_3$$

The corresponding orthogonal and normalized eigenvectors define a basis of another Cartesian coordinate system $\bar{S}_B$ fixed to the rigid body at O_B. Let us call $\bar{S}_B$ the principal coordinate system attached to O_B. The coordinate axes of this coordinate system are called the principal axes of the rigid body for the given point O_B. The term *principal axes* without further specification usually refers to the principal axes that pass through the center of mass of the rigid body. The principal coordinate system attached to the center of mass is denoted by $\bar{S}_{CM}$. Clearly, the inertia tensor is diagonal in the coordinate system $\bar{S}_B$:

$$\bar{I} = \begin{pmatrix} I_1 & 0 & 0 \\ 0 & I_2 & 0 \\ 0 & 0 & I_3 \end{pmatrix} \tag{9.40}$$

If it is found that the inertia tensor is already in diagonal form in a given coordinate system S_B, then the coordinate axes of S_B are the principal axes. Principal axes may coincide with some of the geometrical axes of symmetry, provided that they exist for a rigid body.

Two rigid bodies having the same set of principal moments of inertia are dynamically equivalent with respect to rotation.

Example 9.6

Inspecting Table 9.1 we find that the ellipsoid $x^2/a^2 + y^2/b^2 + z^2/c^2 = 1$ of mass m_{ell} and the rectangular prism of sides $2a$, $2b$, and $2c$ and mass $m_{pr} = (3/5)m_{ell}$ have the same principal moments of inertia. ∎

We shall follow the convention that the basis vectors of $\bar{S}_B$ form a right-handed triad. The transformation of the basis vectors from S_B into $\bar{S}_B$ is, in general, a combination of a rotation and a reflection, which, as we know from Section 2.9, are described by orthogonal matrices. If A denotes the orthogonal matrix that defines the transformation from S_B into $\bar{S}_B$, then the connection between and $\bar{\mathsf{I}}$ is given by

$$\bar{\mathsf{I}} = \mathsf{AIA}^T, \tag{9.41}$$

see (2.89).

It is important to make a clear distinction between the coordinate systems S fixed in space, and S_B, $\bar{S}_B$, S_{CM}, and $\bar{S}_{CM}$ fixed to the rigid body (see Figure 9.3). Note that the coordinate axes of $\bar{S}_B$ and $\bar{S}_{CM}$ are not parallel to each other in general.

Example 9.7

Find the principal moments of inertia and the principal axes for the system described in Example 9.1
    The characteristic equation

$$|\mathsf{I} - \lambda\mathsf{1}| = \begin{vmatrix} 2 - \lambda & 0 & -2 \\ 0 & 6 - \lambda & 0 \\ -2 & 0 & 4 - \lambda \end{vmatrix} = -\lambda^3 + 12\lambda^2 - 40\lambda + 24 = 0$$

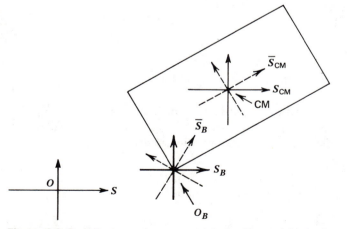

Figure 9.3 Coordinate systems associated with a rigid body. Dashed axes refer to the principal systems. The system S is fixed in space.

has the three roots $3 - \sqrt{5}$, $3 + \sqrt{5}$, and 6. Therefore, $\bar{I}$ is given by

$$\bar{I} = \begin{pmatrix} 3 - \sqrt{5} & 0 & 0 \\ 0 & 3 + \sqrt{5} & 0 \\ 0 & 0 & 6 \end{pmatrix}$$

The eigenvector $X^{(1)}$ belonging to $I_1 = 3 - \sqrt{5}$ is found from

$$\left[I - (3 - \sqrt{5})1 \right] X^{(1)} = 0$$

or, in coordinate form,

$$\begin{aligned}
(\sqrt{5} - 1)x_1 & & -2x_3 &= 0 \\
& (3 + \sqrt{5})x_2 & &= 0 \\
-2x_1 & & +(1 + \sqrt{5})x_3 &= 0
\end{aligned}$$

This system of equations has the solution

$$x_1 = 1, \quad x_2 = 0, \quad x_3 = \tfrac{1}{2}(\sqrt{5} - 1)$$

The coordinates of the normalized eigenvector $X^{(1)}$ are

$$x_1^{(1)} = \frac{1}{\sqrt{\tfrac{1}{2}(5 - \sqrt{5})}}, \quad x_2^{(1)} = 0, \quad x_3^{(1)} = \frac{1}{\sqrt{\tfrac{1}{2}(5 + \sqrt{5})}}$$

Similarly, one obtains

$$x_1^{(2)} = \frac{-1}{\sqrt{\tfrac{1}{2}(5 + \sqrt{5})}}, \quad x_2^{(2)} = 0, \quad x_3^{(2)} = \frac{1}{\sqrt{\tfrac{1}{2}(5 - \sqrt{5})}}$$

and

$$x_1^{(3)} = 0, \quad x_2^{(3)} = -1, \quad x_3^{(3)} = 0$$

The three eigenvectors define a right-handed orthonormal basis of $\bar{S}_B$. The coordinate axes of $\bar{S}_B$ are the principal axes of the rigid body. The reader may verify that (9.41) is satisfied with the transformation matrix

$$A = \begin{pmatrix} \cos \theta & 0 & \sin \theta \\ -\sin \theta & 0 & \cos \theta \\ 0 & -1 & 0 \end{pmatrix}$$

where $\theta = \arctan(x_3^{(1)}/x_1^{(1)}) = 31.7°$ is the angle of rotation in clockwise direction about the x_2-axis. ∎

9.4.7 Angular Momentum and Kinetic Energy

The general structure of the angular momentum $\mathbf{L}$ expressed by (9.25) led in a natural way to the definition of the inertia tensor (9.26). Because the inertia tensor is diagonal in the principal system, the coordinates of $\mathbf{L}$ in $\bar{S}_B$ simplify to

$$\bar{L}_k = \sum_{l=1}^{3} I_k \delta_{kl} \bar{\omega}_l = I_k \bar{\omega}_k \qquad (9.42)$$

or

$$\bar{L}_1 = I_1 \bar{\omega}_1, \quad \bar{L}_2 = I_2 \bar{\omega}_2, \quad \bar{L}_3 = I_3 \bar{\omega}_3 \qquad (9.43)$$

We conclude that the angular momentum vector is parallel to the angular velocity vector if the axis of rotation coincides with a principal axis. In particular, $\mathbf{L}$ is always parallel to $\boldsymbol{\omega}$ for bodies whose inertia tensor is a multiple of the unit matrix, that is, if $I_1 = I_2 = I_3$.

Let us now calculate the rotational kinetic energy of a rigid body in terms of $\mathbf{I}$ and $\boldsymbol{\omega}$. Relative to a fixed point O_B the velocity vector $\dot{\mathbf{r}}$ is given by $\boldsymbol{\omega} \times \mathbf{r}$, which we substitute into (9.7). The following manipulations are similar to those for the derivation of the angular momentum with the summation over particles replaced by integrals. We have

$$T_{rot} = \int_v \frac{1}{2} \dot{\mathbf{r}} \cdot \dot{\mathbf{r}} \rho \, dV$$

$$= \int_v \frac{1}{2} (\boldsymbol{\omega} \times \mathbf{r}) \cdot (\boldsymbol{\omega} \times \mathbf{r}) \rho \, dV$$

$$= \int_v \frac{1}{2} \left[r^2 \omega^2 - (\boldsymbol{\omega} \cdot \mathbf{r})(\boldsymbol{\omega} \cdot \mathbf{r}) \right] \rho \, dV$$

$$= \sum_{k=1}^{3} \sum_{l=1}^{3} \frac{1}{2} \omega_k \left[\int_v (r^2 \delta_{kl} - x_k x_l) \rho \, dV \right] \omega_l$$

$$= \sum_{k=1}^{3} \sum_{l=1}^{3} \frac{1}{2} \omega_k I_{kl} \omega_l \qquad (9.44)$$

In the principal coordinate system $\bar{S}_B$ the kinetic energy becomes

$$T_{rot} = \frac{1}{2} \sum_{k=1}^{3} I_k \bar{\omega}_k^2 \qquad (9.45)$$

If the center of mass is chosen as the reference point, the rotational kinetic energy

$$T'_{rot} = \frac{1}{2} \sum_{k=1}^{3} I'_k \bar{\omega}_k^2$$

is just the term T' appearing in (7.44) when applied to rigid bodies.

Occasionally it is useful to relate the kinetic energy to the angular momentum. Comparing Eq. 9.25 with Eq. 9.40, we obtain

$$T = \frac{1}{2} \sum_{k=1}^{3} \omega_k L_k$$

$$= \frac{1}{2}\boldsymbol{\omega} \cdot \mathbf{L} \tag{9.46}$$

9.4.8 The Inertia Ellipsoid

Let x_1, x_2, and x_3 be the coordinates of a position vector in the arbitrary Cartesian coordinate system S_B, and let $\bar{x}_1$, $\bar{x}_2$, and $\bar{x}_3$ be the coordinates of the same vector in the principal system $\bar{S}_B$. The x_k and the $\bar{x}_l$ are then related to each other by an orthogonal transformation,

$$\overline{\mathbf{X}} = \mathbf{A}\mathbf{X} \tag{9.47}$$

where $\mathbf{A}$ is an orthogonal matrix.

The equation

$$\mathbf{X}^T \mathbf{I} \mathbf{X} = 1, \tag{9.48}$$

or in explicit form

$$I_{11}x_1^2 + I_{22}x_2^2 + I_{33}x_3^2 + 2I_{12}x_1x_2 + 2I_{13}x_1x_3 + 2I_{23}x_2x_3 = 1,$$

defines a surface in the three-dimensional space. Using Eqs. 9.41 and 9.47, and remembering that for orthogonal matrices $\mathbf{A}^T\mathbf{A} = 1$ holds, we can write (9.48) in the form

$$\mathbf{X}^T 1 \mathbf{I} 1 \mathbf{X} = \mathbf{X}^T \mathbf{A}^T \mathbf{A} \mathbf{I} \mathbf{A}^T \mathbf{A} \mathbf{X}$$

$$= (\mathbf{A}\mathbf{X})^T (\mathbf{A}\mathbf{I}\mathbf{A}^T)(\mathbf{A}\mathbf{X})$$

$$= \overline{\mathbf{X}}^T \overline{\mathbf{I}} \overline{\mathbf{X}} = 1$$

The last line reads explicitly

$$I_1\bar{x}_1^2 + I_2\bar{x}_2^2 + I_3\bar{x}_3^2 = 1 \tag{9.49}$$

Because the principal moments of inertia are not negative, this surface must be an ellipsoid, called the inertia ellipsoid. The semiaxes of the ellipsoid are $I_1^{-1/2}$, $I_2^{-1/2}$, and $I_3^{-1/2}$, respectively, and they are arranged in the order of decreasing magnitude. The inertia ellipsoid enables us to visualize the inertia tensor geometrically and to draw several conclusions about the inertial properties of the rigid body without using much mathematics. Remember that the inertia ellipsoid is fixed to the rigid body and not to the coordinate system used.

Let us consider a rigid body whose principal moments of inertia I_k (in $\bar{S}_B$) are all different. Because the I_k are directly related to the semimajor axes of the

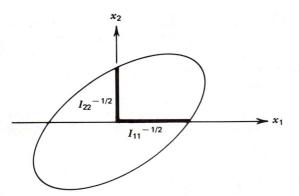

Figure 9.4 *a* Inertia ellipsoid in S_B.

inertia ellipsoid, it follows that I_1 is the smallest, and that I_3 is the largest possible moment of inertia of the rigid body about all the axes passing through the same point. Moreover, these axes of rotation are perpendicular to each other. In S_B, the moments of inertia I_{kk} are related to the coordinate x_k of the intersection of the x_k-axis with the inertia ellipsoid by

$$x_k = \frac{1}{\sqrt{I_{kk}}},\qquad(9.50)$$

(see Problem 9.8). It follows then that I_{kk} does not, in general, assume an extremum (see Figures 9.4a and 9.4b). If the I_k are all different from each other, the principal axes of the rigid body and, therefore, the semiaxes of the inertia ellipsoid are uniquely determined.

If two of the I_k are equal to each other, that is, in the case of degeneracy of the eigenvalues of I, the inertia ellipsoid possesses rotational symmetry about the third axis. In the case $I_1 = I_2$, the intersection of the inertia ellipsoid with the x_1-x_2 plane is a circle. Then there are infinitely many pairs of orthogonal coordinate axes in the x_1-x_2 plane that may serve as principal axes, and the moments of inertia are the same for all those axes. This explains the result

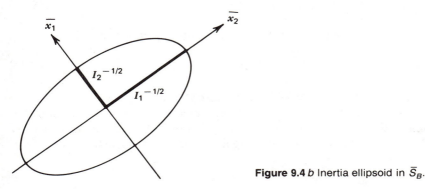

Figure 9.4 *b* Inertia ellipsoid in $\bar{S}_B$.

obtained in Example 9.4 for the moment of inertia of a square lamina. If $I_1 = I_2 = I_3$, the inertia ellipsoid reduces to a sphere, and the moments of inertia about any axis passing through the same point O_B are equal to each other. A cube of homogeneous mass distribution, for example, possesses this property of rotational isotropy about the center of mass.

A rigid body for which $I_1 = I_2 = I_3$ is called a spherical top. A body for which $I_1 = I_2 \neq I_3$ or $I_1 \neq I_2 = I_3$ is called a symmetrical top. If $I_1 = 0$ and $I_2 = I_3$, the body is called a rotor. If all principal moments of inertia are different from each other the rigid body is termed an asymmetrical top.

9.5 ROTATION ABOUT A FIXED AXIS

9.5.1 The Equation of Motion

In this and the following sections we shall investigate the dynamical equations for rotation in more detail.

We begin with the simplest case, namely, the rotation of the rigid body about a fixed axis. This is the one-dimensional equivalent of the translational motion of a mass point in one direction. If we call the axis of rotation the z-axis, then the angular velocity vector is given by

$$\boldsymbol{\omega} = (0, 0, \omega_z)$$

If the axis of rotation coincides with a principal axis of the rigid body, the coordinates of the angular momentum vector are by (9.43) equal to

$$L_x = 0, \quad L_y = 0, \quad L_z = I_z \omega_z \tag{9.51}$$

Because I_z is a constant, we have

$$\dot{L}_z = I_z \dot{\omega}_z$$

Using the torque in the form of (9.19) we obtain the equation of motion

$$I_z \dot{\omega}_z = (\mathbf{R} \times \dot{\mathbf{P}})_z + N_z'$$
$$= (\mathbf{R} \times \mathbf{F})_z + N_z' \tag{9.52}$$

which for rotations about an axis through the center of mass simplifies to

$$I_z' \dot{\omega}_z = N_z' \tag{9.53}$$

by the parallel axis theorem (9.38), and because $\mathbf{R} = 0$.

We have already stated that the angular momentum vector is not parallel to the angular velocity vector if the axis of rotation does not coincide with a principal axis. The coordinates of the angular momentum are then

$$L_x = I_{xz}\omega_z, \quad L_y = I_{yz}\omega_z, \quad L_z = I_{zz}\omega_z$$

In this case, the angular momentum perpendicular to the z-axis is transmitted to the support of the rigid body. If the mass of the support is very large, the axis of rotation will not change appreciably. If the mass of the support is small, the transmitted angular momentum causes the whole system to wobble, and the z-axis no longer remains the axis of rotation. In order to compel the system to rotate about the fixed z-axis, torques of constraint must act on the rigid body perpendicular to the axis of rotation. The equations of motion (9.52) and (9.53) with I_z replaced by I_{zz} remain unaffected.

Almost everyone has felt the vibrations produced by unbalanced tires on a car. These vibrations, together with the torques of constraints on the wheel, may cause damage to the wheel bearings. These unwanted effects can be eliminated by balancing the tire. One distinguishes two types of balancing: Dynamic balancing refers to the elimination of the products of inertia in order to get rid of the unwanted torques of constraint. Static balancing refers to the alignment of the axis of rotation through the center of mass in order to eliminate the torque $\mathbf{R} \times \dot{\mathbf{P}} = \mathbf{R} \times \mathbf{F}$. Balancing is achieved by adding masses to the wheel at certain positions.

9.5.2 The Compound Pendulum

The oscillation of a physical pendulum is a standard example for the motion of a rigid body about a fixed axis.

The physical or compound pendulum is a rigid body that is free to oscillate about a fixed axis under the influence of the gravitational force (see Figure 9.5). Let h be the distance from the axis of rotation through O_B to the center of mass. The moment of inertia about the point O_B is by the parallel axis theorem (9.38) equal to

$$I_{zz} = Mh^2 + I'_{zz}$$

Because $\mathbf{N}' = 0$ for gravitational forces (see Eq. 9.23), the equation of motion

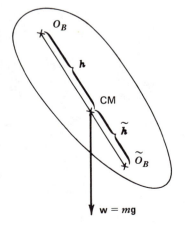

Figure 9.5 The compound pendulum.

(9.52) reads

$$I_{zz}\dot{\omega}_z = (\mathbf{h} \times \mathbf{w})_z$$

or

$$\left(Mh^2 + I'_{zz} \right)\ddot{\theta} = -hMg \sin \theta \qquad (9.54)$$

For small amplitudes of oscillation we can use the approximation $\sin \theta \approx \theta$, and Eq. 9.54 can be written in the form

$$\ddot{\theta} + \frac{hMg}{Mh^2 + I'_{zz}}\theta = 0$$

The frequency of oscillation is, by (5.5) and (5.19), equal to

$$f = \frac{1}{2\pi}\sqrt{\frac{hMg}{Mh^2 + I'_{zz}}} = \frac{1}{2\pi}\sqrt{\frac{g}{h + \dfrac{I'_{zz}}{Mh}}} \qquad (9.55)$$

The physical pendulum behaves like a simple pendulum of length

$$l = h + \frac{1}{h}\frac{I'_{zz}}{M}, \qquad (9.56)$$

(see Section 5.6). Remembering that

$$k_{\text{CM}} = \sqrt{\frac{I'_{zz}}{M}}$$

is the radius of gyration about the center of mass, we can write (9.56) as

$$l = h + \frac{k_{\text{CM}}^2}{h} \qquad (9.57)$$

It is possible to find another point $\tilde{O}_B$ along the axis joining the points O_B and the center of mass such that the frequency of oscillation assumes the same value as that given by (8.55). A physical pendulum that is designed to oscillate about the two points O_B and $\tilde{O}_B$ is called Kater's reversible pendulum. The distance $\tilde{h}$ from $\tilde{O}_B$ to the center of mass must also satisfy the condition (9.57) with h replaced by $\tilde{h}$,

$$l = \tilde{h} + \frac{k_{\text{CM}}^2}{\tilde{h}} \qquad (9.58)$$

It is then easy to show that

$$h + \tilde{h} = l \qquad (9.59)$$

218   RIGID BODY MECHANICS

and

$$h\tilde{h} = k^2_{CM},$$ (9.60)

(see Problem 9.13). Because h, $\tilde{h}$, and f can be measured with high precision, Kater's pendulum is a very precise instrument for determining the value of the gravitational acceleration g. $\tilde{O}_B$ is called the center of oscillation for the point O_B, and vice versa.

9.5.3 The Center of Percussion

Equation 9.60 is also obtained in an entirely different problem. We consider a rigid body that is initially at rest. We wish to find that point $\tilde{O}_B$ of a rigid body where an impulsive force must act such that a given point O_B does not experience a linear acceleration, at least during a short time interval (see Figure 9.6). Under the given conditions, the point O_B is the reference point for the rotation. Relative to O_B the equation of rotational motion reads

$$(I' + Mh^2)\dot{\omega} = (h + \tilde{h})F$$ (9.61)

We also know that the net force accelerates the total mass located at the center of mass, thus

$$Ma_{CM} = F$$ (9.62)

The center of mass acceleration a_{CM} in turn is related to the angular acceleration by $a_{CM} = h\dot{\omega}$. Multiplying Eq. 9.62 by h we obtain

$$Mh^2\dot{\omega} = hF$$ (9.63)

Subtracting this equation from (9.61) we find

$$I'\dot{\omega} = \tilde{h}F$$ (9.64)

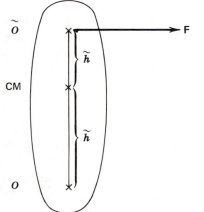

Figure 9.6 Center of percussion.

Finally, division of (9.64) by (9.63) yields

$$\frac{I'}{M}\frac{1}{h^2} = \frac{\tilde{h}}{h},$$

or $h\tilde{h} = k_{\mathrm{CM}}^2$, which is Eq. 9.60.

$\tilde{O}_B$ is therefore also called the center of percussion relative to O_B, and vice versa. To eliminate any jerk of the hands of a baseball player, the baseball should be struck at the center of percussion of the bat relative to the position of the hands. The area around the center of percussion of a tennis racket is known as the "sweet spot."

9.6 ROTATION ABOUT A FIXED POINT

9.6.1 Euler's Equations

In this section we shall investigate the rotation of a rigid body about a given point O_B which is held fixed in space at all times. The rigid body does not then perform a translational motion. We are interested in two special cases: the rotation about the center of mass (e.g., the rotating earth, or a rotating box thrown into the air), and the rotation about a point lying on the symmetry axis of a cylindrically symmetric object (e.g., a spinning top). It is simplest to let the origin O of the space system S coincide with the origin O_B of the body system. Clearly, as the rigid body rotates in space, the coordinate system S_B rotates relative to S, and if the coordinate systems S and S_B are both orthogonal systems, the coordinates of a given vector in S are related to those in S_B by an orthogonal transformation. Because the rigid body is free to rotate about three arbitrary axes, for example, the x-, y-, and z-axes, the number of degrees of freedom is three.

We wish to write the equation of motion

$$\dot{\mathbf{L}} = \mathbf{N},$$

Eq. 9.9, in a form that contains the moments of inertia explicitly. We must not forget that Eq. 9.9 holds in the space system S (inertial frame), and that both $\mathbf{L}$ and $\mathbf{N}$ are defined relative to the point O_B. Equation 9.9 reads in coordinate form

$$\dot{L}_k = \left(\frac{dL_k}{dt}\right)_S = N_k \tag{9.65}$$

where the subscript S is used to indicate that the derivative must be taken in the space system.

It is not advantageous to express the coordinates of the angular momentum in the space system because the inertia tensor is not constant in S. However, the inertia tensor is constant in any body system such that the time derivative of L_k in

S_B is equal to

$$\left(\frac{dL_k}{dt}\right)_B = \sum_{l=1}^{3} I_{kl} \left(\frac{d\omega_l}{dt}\right)_B \qquad (9.66)$$

The connection between the time derivative of any vector $\mathbf{q}$ relative to S and the time derivative of the same vector relative to S_B is given by (3.33):

$$\left(\frac{d\mathbf{q}}{dt}\right)_S = \left(\frac{d\mathbf{q}}{dt}\right)_B + \boldsymbol{\omega} \times \mathbf{q} \qquad (9.67)$$

Setting $\mathbf{q} = \boldsymbol{\omega}$ we have

$$\dot{\boldsymbol{\omega}} = \left(\frac{d\boldsymbol{\omega}}{dt}\right)_S = \left(\frac{d\boldsymbol{\omega}}{dt}\right)_B \qquad (9.68)$$

The angular acceleration is the same in both the space system and the body system. Applying Eq. 9.64 to the angular momentum, and combining (9.65), (9.66), and (9.68) we obtain for the time derivative of L_1 the expression

$$\left(\frac{dL_1}{dt}\right)_S = \left(\frac{dL_1}{dt}\right)_B + \omega_2 L_3 - \omega_3 L_2$$

$$= \sum_{l=1}^{3} I_{1l}\dot{\omega}_l + \omega_2 L_3 - \omega_3 L_2 = N_1 \qquad (9.69)$$

The last equation reduces to a simpler form in the principal system $\bar{S}_B$. Because $I_{kl} = I_k \delta_{kl}$, and $L_k = I_k \omega_k$ by (9.42) (the bars are omitted for simplicity), we find

$$I_1 \dot{\omega}_1 + \omega_2 \omega_3 (I_3 - I_2) = N_1$$
$$I_2 \dot{\omega}_2 + \omega_3 \omega_1 (I_1 - I_3) = N_2$$
$$I_3 \dot{\omega}_3 + \omega_1 \omega_2 (I_2 - I_1) = N_3 \qquad (9.70)$$

where the equations for the second and the third component have been obtained from the first by cyclic permutation of the indices: $1 \rightarrow 2 \rightarrow 3 \rightarrow 1$. Equations 9.70 are called Euler's equations. They form a set of three nonlinear differential equations for the angles of rotation. Fortunately, for some important situations, exact solutions of Euler's equations exist, although they may contain special mathematical functions (elliptic integrals).

It is important to remember that the components of all physical quantities which appear in Euler's equations are formulated in the principal body system $\bar{S}_B$, which is attached to the point O_B fixed in space. The torque $\mathbf{N}$ is defined relative to the point O_B. Although the coordinates of $\mathbf{N}$ are expressed relative to S_B, the torque is produced by forces originating from some source in the space system S.

The coordinates of the angular velocity vector relative to S are obtained from the coordinates relative to $\bar{S}_B$ by application of a coordinate transformation.

Euler's equations assume a simpler form if the rigid body possesses symmetries.

If $I_1 = I_2 = I_3 = I$, Euler's equations are

$$I\dot{\omega}_1 = N_1,$$

$$I\dot{\omega}_2 = N_2,$$

$$I\dot{\omega}_3 = N_3 \tag{9.71}$$

These equations are formally equal to the dynamical equations for translational motion of a point particle. This is easily explained: The mass of a particle is (always) a scalar, whereas the rotational inertia is a tensor; but for the symmetric top the inertia tensor reduces to a multiple of the unit matrix, which may be treated like a scalar.

If $I_1 = I_2 \neq I_3$, Euler's equations are

$$I_1\dot{\omega}_1 + \omega_2\omega_3(I_3 - I_1) = N_1,$$

$$I_1\dot{\omega}_2 - \omega_3\omega_1(I_3 - I_1) = N_2,$$

$$I_3\dot{\omega}_3 = N_3 \tag{9.72}$$

Similar equations are obtained if $I_1 \neq I_2 = I_3$.

9.6.2 Conservation Theorems for Rigid Body Rotation

We mentioned in the introduction to this chapter that the angular momentum $\mathbf{L}$ is conserved if the rotational impulse

$$\mathcal{G}_{rot} = \int_{t_1}^{t_2} \mathbf{N} \, dt \tag{9.73}$$

vanishes. Also, if the work

$$W_{rot} = \int_{t_1}^{t_2} \mathbf{N} \cdot \boldsymbol{\omega} \, dt \tag{9.74}$$

vanishes, the rotational kinetic energy T_{rot} is conserved by the work energy theorem (see Problem 9.20). Trivially then, if the rigid body is not acted upon by any torque, $\mathbf{N} = 0$, both $\mathbf{L}$ and T_{rot} are constants of the motion.

Although conservation of linear momentum of a mass point is equivalent to the constancy of the linear velocity, conservation of angular momentum does not, in general, imply that the angular velocity is also a constant.

Let us investigate more closely the angular velocity of a torque free rigid body. We begin by writing Euler's equations (9.70) in the form

$$\dot{\omega}_1 = -\omega_2\omega_3\frac{I_3 - I_2}{I_1}$$

$$\dot{\omega}_2 = -\omega_3\omega_1\frac{I_1 - I_3}{I_2}$$

$$\dot{\omega}_3 = -\omega_1\omega_2\frac{I_2 - I_1}{I_3} \tag{9.75}$$

Multiplying each equation by the corresponding angular velocity coordinate, and adding the resulting equations, we obtain

$$\dot{\omega}_1\omega_1 + \dot{\omega}_2\omega_2 + \dot{\omega}_3\omega_3 = -\omega_1\omega_2\omega_3\left(\frac{I_3 - I_2}{I_1} + \frac{I_1 - I_3}{I_2} + \frac{I_2 - I_1}{I_3}\right) \tag{9.76}$$

The left side of this equation is equal to $(1/2)\, d\omega^2/dt$. If $I_1 \neq I_2 \neq I_3$, the right side does not identically vanish, and the angular speed ω and the angular velocity $\boldsymbol{\omega}$ are not constant. If either $I_1 = I_2$ or $I_2 = I_3$, the term in parentheses is zero and the angular speed is constant. However, the angular velocity is not constant because $\boldsymbol{\omega}$ is not proportional to $\mathbf{L}$. If $I_1 = I_2 = I_3$, Eqs. 9.75 reduce to $\dot{\omega}_k = 0$, which implies that $\boldsymbol{\omega} = $ constant. These results are summarized in Table 9.2.

9.6.3 Stability of Torque-Free Rotation

We consider a rigid body that is not acted upon by any forces or torques. In the absence of forces the rigid body rotates about the center of mass, which remains at rest relative to some inertial frame. Euler's equations must then be set up in the principal center of mass frame $\bar{S}_{CM}$.

We assume that the principal moments of inertia are different from each other, $I_1 < I_2 < I_3$. Introducing the positive constants

$$A_1 = \frac{I_3 - I_2}{I_1}, \quad A_2 = \frac{I_3 - I_1}{I_2}, \quad A_3 = \frac{I_2 - I_1}{I_3} \tag{9.77}$$

Table 9.2 Conserved Quantities of a Torque-Free Rigid Body

I	T_{rot} [in S]	$\mathbf{L}$ [in S]	ω^2 [in $\bar{S}_{CM}$]	$\boldsymbol{\omega}$ [in $\bar{S}_{CM}$]	Stability
$I_1 < I_2 < I_3$	Constant	Constant	$\neq$ Constant	$\neq$ Constant	Stable about the 1- and 3-axis
$\left.\begin{array}{l} I_1 = I_2 < I_3 \\ I_1 < I_2 = I_3 \end{array}\right\}$	Constant	Constant	Constant	$\neq$ Constant	Stable about the 3-axis Stable about the 1-axis
$I_1 = I_2 = I_3$	Constant	Constant	Constant	Constant	Neutral about all axes

we can write Euler's equations (9.75) for torque-free rotation in the form

$$\dot{\omega}_1 + \omega_2\omega_3 A_1 = 0 \tag{9.78}$$

$$\dot{\omega}_2 - \omega_3\omega_1 A_2 = 0 \tag{9.79}$$

$$\dot{\omega}_3 + \omega_1\omega_2 A_3 = 0 \tag{9.80}$$

Suppose that initially (at $t = 0$) $\omega_1 \gg \omega_2 \approx \omega_3$. The product $\omega_2\omega_3$ is then small of second order such that from (9.78) follows $\omega_1 = $ constant. Taking the time derivative of (9.79), and eliminating the term $\dot{\omega}_3$ using (9.80), we obtain

$$\ddot{\omega}_2 + \left(\omega_1^2 A_2 A_3\right)\omega_2 = 0 \tag{9.81}$$

Interchanging the operations on Eqs. 9.79 and 9.80, we arrive at

$$\ddot{\omega}_3 + \left(\omega_1^2 A_2 A_3\right)\omega_3 = 0 \tag{9.82}$$

The general solutions of (9.81) and (9.82) are of the form

$$A \sin\left(\omega_1\sqrt{A_2 A_3}\, t + \delta\right)$$

The small angular velocity components ω_2 and ω_3 are bounded and change periodically with the angular frequency $\omega_1\sqrt{A_2 A_3}$.

If initially $\omega_3 \gg \omega_1 \approx \omega_2$, similar conclusions can be drawn for the small angular velocity components ω_1 and ω_2. Both ω_1 and ω_2 are bounded and vary with the angular frequency $\omega_3\sqrt{A_1 A_2}$.

An entirely different situation arises if initially $\omega_2 \gg \omega_1 \approx \omega_3$. We then have $\omega_2 = $ constant, but the elimination process leads to

$$\ddot{\omega}_1 - \left(\omega_2^2 A_1 A_3\right)\omega_1 = 0$$

and

$$\ddot{\omega}_3 - \left(\omega_2^2 A_1 A_3\right)\omega_3 = 0$$

The general solution of these equations is

$$Ae^{+\omega_2\sqrt{A_1 A_3}\, t} + Be^{-\omega_2\sqrt{A_1 A_3}\, t}$$

Because the first term is unbounded as t increases, the motion is unstable.

We have derived the following result: Rotational motion of a rigid body not experiencing any torque is stable only about the two principal axes for which the moment of inertia is a minimum or a maximum.

It is left as an exercise to show that for $I_1 = I_2 < I_3$ and for $I_1 < I_2 = I_3$ only one stable axis of rotation exists (see Problem 9.21).

9.6.4 Torque-Free Motion of a Symmetrical Top

We shall now investigate the torque-free motion of a rigid body with axial symmetry. We consider the case $I_1 = I_2 < I_3$, which is typical for a gyroscope. Euler's equations (9.72) are then

$$\dot{\omega}_1 + \omega_2\omega_3\frac{I_3 - I_1}{I_1} = 0$$

$$\dot{\omega}_2 - \omega_1\omega_3\frac{I_3 - I_1}{I_1} = 0$$

$$\dot{\omega}_3 = 0 \qquad (9.83)$$

in the frame $\bar{S}_{CM}$. The last equation implies that

$$\omega_3 = \text{constant}$$

Introducing the constant

$$\Omega = \omega_3\frac{I_3 - I_1}{I_1} \qquad (9.84)$$

we can write the equations for ω_1 and ω_2 in the simpler form

$$\dot{\omega}_1 + \Omega\omega_2 = 0, \qquad (9.85a)$$

$$\dot{\omega}_2 - \Omega\omega_1 = 0 \qquad (9.85b)$$

Note that these equations are no longer nonlinear. Decoupling Eqs. 9.85 by differentiation and elimination of the first order derivative, we obtain the differential equations

$$\ddot{\omega}_1 + \Omega^2\omega_1 = 0, \qquad (9.86)$$

$$\ddot{\omega}_2 + \Omega^2\omega_2 = 0 \qquad (9.87)$$

which are of the simple harmonic oscillator type. The solutions of (9.86) and (9.87),

$$\omega_1 = \omega_0\sin(\Omega t + \delta)$$

and

$$\omega_2 = \omega_0'\cos(\Omega t + \delta')$$

are by (9.85) consistent with each other only if

$$\omega_0 = \omega_0' \qquad \text{and} \qquad \delta = \delta'$$

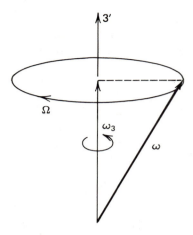

Figure 9.7 Precession in the body system.

Therefore, we have, setting $\delta = 0$ for simplicity,

$$\omega_1 = \omega_0 \sin \Omega t,$$

$$\omega_2 = \omega_0 \cos \Omega t, \tag{9.88}$$

and

$$\omega_1^2 + \omega_2^2 = \omega_0^2 \tag{9.89}$$

It was shown in Section 6.2 that also

$$\omega^2 = \omega_1^2 + \omega_2^2 + \omega_3^2 = \text{constant} \tag{9.90}$$

The vector $\omega_1 \mathbf{e}_1' + \omega_2 \mathbf{e}_2'$, where $\mathbf{e}_1'$ and $\mathbf{e}_2'$ are basis vectors in $\bar{S}_{\text{CM}}$, rotates about the 3'-axis with constant angular frequency Ω in the direction opposite to ω_3. Because ω_3 is constant, the vector ω precesses about the 3'-axis of the rigid body (see Figure 9.7). Ω is therefore called the angular frequency of precession. By its definition (9.84), Ω may become very small if $I_1 \approx I_3$. Clearly, if $I_1 = I_2 = I_3$, no precession occurs, and the angular velocity vector is a constant.

The integration constants ω_0 and ω_3 may be obtained from the initial conditions expressed in terms of the conserved angular momentum

$$L^2 = I_1^2 \omega_0^2 + I_3^2 \omega_3^2$$

and the conserved kinetic energy

$$T_{rot} = \tfrac{1}{2} I_1 \omega_0^2 + \tfrac{1}{2} I_3 \omega_3^2$$

Figure 9.8 Precession in the space system.

Writing Eq. 9.46 in the form

$$T = \tfrac{1}{2}\omega L \cos\sphericalangle(\omega, \mathbf{L})$$

we see that the angle $\theta = \sphericalangle(\omega, \mathbf{L})$ is a constant of the motion. Because the angular momentum vector $\mathbf{L}$ is fixed in the space system S, the angular velocity vector ω traces out a circular cone about $\mathbf{L}$ (see Figure 9.8). Remember that the coordinates of ω were calculated in the body frame $\bar{S}_{\mathrm{CM}}$, whereas the conservation of $\mathbf{L}$ is expressed in the space system S. The coordinates of $\mathbf{L}$ in $\bar{S}_{\mathrm{CM}}$ are

$$I_1\omega_0\sin\Omega t, \quad I_1\omega_0\cos\Omega t, \quad I_3\omega_3,$$

and only $L_{3(\text{body})}$ is a constant. It follows that the angle between the body 3-axis and the angular momentum vector is a constant and that, therefore, the 3-axis moves on a circular cone about $\mathbf{L}$. The 3-axis, as the symmetry axis of the body, can easily be identified when the body is rotating, and the line from the center of mass to the center of the circle described by a point on the 3′-axis defines the direction of the angular momentum.

In gyroscopes, the angular momentum vector is usually directed along the symmetry axis (3′-axis) such that

$$\mathbf{L} = I_3\omega_3\mathbf{e}_3',$$

and, in the absence of any torque, the symmetry axis remains fixed in an inertial frame. For this reason gyroscopes find numerous applications in technology, for examples as stabilizers on rockets and space platforms and in the form of gyrocompasses in navigation and guidance systems. There are indications that optical gyroscopes[1] using lasers will soon replace mechanical gyroscopes in navigation systems.

[1]See E. Hecht and A. Zajac: *Optics*, Addison-Wesley, Reading, Mass: 1979; "Sensitive Fiber-Optics Gyroscopes," *Physics Today*, October 1981, p. 20.

9.6.5 Eulerian Angles

The angles θ_1, θ_2, and θ_3 describing the rotation of a rigid body about the three Cartesian coordinate axes of the space system S are not always convenient generalized coordinates. Often, the rotation of a rigid body is better understood in terms of the Eulerian angles.

We have seen in Section 3.3 that any rotation can be represented by an orthogonal matrix with determinant $+1$. Conversely, any orthogonal matrix with determinant $+1$ describes a rotation. Because the product of two orthogonal matrices is another orthogonal matrix, any number of rotations performed in succession is equivalent to one rotation about some fixed axis. In particular, the three independent rotations $R_1(\theta_1)$, $R_2(\theta_2)$, and $R_3(\theta_3)$ about the x_1-, x_2-, and x_3-axes, respectively, where θ_1, θ_2, and θ_3 are the angles of rotation, are equivalent to one rotation

$$R = R_3 R_2 R_1$$

Because matrix multiplication need not be commutative, $AB \neq BA$, the inverse process of decomposing an arbitrary rotation R into rotations about the coordinate axes is not unique unless the order of the rotations is prescribed.

We assume that initially the Cartesian basis vectors $\{e_1, e_2, e_3\}$ of the space system S coincide with the Cartesian basis vectors $\{e'_1, e'_2, e'_3\}$ of the body system S_B. The Eulerian angles are now defined by the following operations performed in the indicated order (see Figures 9.9a, 9.9b, and 9.9c).

1. The precession angle ϕ is the angle of rotation performed counterclockwise about the vector e_3. The new basis vectors of S_B are $\{e_\xi, e_{\eta_1}, e_3\}$. The ξ-axis is called the line of nodes.

2. The nutation angle θ is the angle of rotation performed counterclockwise about the vector e_ξ. The basis vectors of S_B are transformed into the set $\{e_\xi, e_{\eta_2}, e'_3\}$.

3. The (body) angle ψ is the angle of rotation performed counterclockwise about the vector e'_3. The final set of basis vectors of S_B is $\{e'_1, e'_2, e'_3\}$.

The rotation of the basis vectors (or coordinate axes) is equivalent to a coordinate transformation. Therefore, using Eqs. 3.18 and observing the directions of the axes of rotation, we can write the rotation matrices associated with the angles ϕ, θ, and ψ as

$$R_\phi = \begin{pmatrix} \cos\phi & \sin\phi & 0 \\ -\sin\phi & \cos\phi & 0 \\ 0 & 0 & 1 \end{pmatrix} \tag{9.91}$$

$$R_\theta = \begin{pmatrix} 1 & 0 & 0 \\ 0 & \cos\theta & \sin\theta \\ 0 & -\sin\theta & \cos\theta \end{pmatrix} \tag{9.92}$$

$$R_\psi = \begin{pmatrix} \cos\psi & \sin\psi & 0 \\ -\sin\psi & \cos\psi & 0 \\ 0 & 0 & 1 \end{pmatrix} \tag{9.93}$$

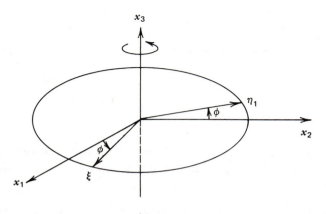

(a)

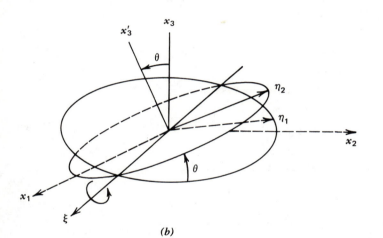

(b)

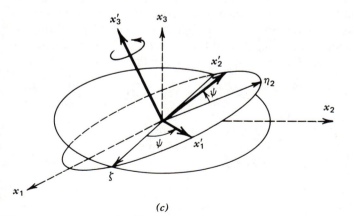

(c)

Figure 9.9 *a*, *b*, *c* Eulerian angles.

The matrix R for the combined rotation is then

$$R = R_\psi R_\theta R_\phi \tag{9.94}$$

or

$$R =$$

$$\begin{pmatrix} \cos\phi\cos\psi - \sin\phi\cos\theta\sin\psi & \sin\phi\cos\psi + \cos\phi\cos\theta\sin\psi & \sin\theta\sin\psi \\ -\cos\phi\sin\psi - \sin\phi\cos\theta\cos\psi & -\sin\phi\sin\psi + \cos\phi\cos\theta\cos\psi & \sin\theta\cos\psi \\ \sin\phi\sin\theta & -\cos\phi\sin\theta & \cos\theta \end{pmatrix}$$

R transforms the coordinates of a vector in S into the coordinates of the same vector in S_B.

By (3.28) the angular velocity tensors corresponding to the point transformations R_ϕ^T, R_θ^T, and R_ψ^T are given by

$$\Omega_\phi = \begin{pmatrix} 0 & -1 & 0 \\ 1 & 0 & 0 \\ 0 & 0 & 0 \end{pmatrix} \dot\phi \tag{9.95}$$

where $\dot\phi$ is the angular velocity about the 3-axis,

$$\Omega_\theta = \begin{pmatrix} 0 & 0 & 0 \\ 0 & 0 & -1 \\ 0 & 0 & 0 \end{pmatrix} \dot\theta \tag{9.96}$$

where $\dot\theta$ is the angular velocity about the ξ-axis, and

$$\Omega_\psi = \begin{pmatrix} 0 & -1 & 0 \\ 1 & 0 & 0 \\ 0 & 0 & 0 \end{pmatrix} \dot\psi \tag{9.97}$$

where $\dot\psi$ is the angular velocity about the 3'-axis.

We wish to find the three independent (Cartesian) components of the angular velocity tensor Ω in the body system S_B in terms of the angular velocities $\dot\phi$, $\dot\theta$, and $\dot\psi$. Ω_ψ is already represented in S_B. Ω_θ must be rotated by the angle θ, whereas Ω_ϕ must be rotated by the angles θ and ϕ about the corresponding axes in this order. Applying the transformation law (2.89) we obtain the result

$$\Omega_B = R_\psi \left(R_\theta \Omega_\phi R_\theta^T \right) R_\psi^T + R_\psi \Omega_\theta R_\psi^T + \Omega_\psi \tag{9.98}$$

or

$$\Omega_B = \begin{pmatrix} 0 & -\cos\theta & \sin\theta\cos\psi \\ \cos\theta & 0 & -\sin\theta\sin\psi \\ -\sin\theta\cos\psi & \sin\theta\sin\psi & 0 \end{pmatrix} \dot\phi$$

$$+ \begin{pmatrix} 0 & 0 & -\sin\psi \\ 0 & 0 & -\cos\psi \\ \sin\psi & \cos\psi & 0 \end{pmatrix} \dot\theta + \begin{pmatrix} 0 & -1 & 0 \\ 1 & 0 & 0 \\ 0 & 0 & 0 \end{pmatrix} \dot\psi$$

from which we identify the components of the angular velocity vector $\boldsymbol{\omega}$ in S_B as

$$\omega_1 = \Omega_{B_{32}} = \sin\theta \sin\psi \dot{\phi} + \cos\psi \dot{\theta}$$

$$\omega_2 = \Omega_{B_{13}} = \sin\theta \cos\psi \dot{\phi} - \sin\psi \dot{\theta}$$

$$\omega_3 = \Omega_{B_{21}} = \cos\theta \dot{\phi} + \dot{\psi} \tag{9.99}$$

It is left as an exercise to show that the components of the angular velocity vector in the space system S are given by

$$\omega_1 = \cos\phi \dot{\theta} + \sin\phi \sin\theta \, \dot{\psi}$$

$$\omega_2 = \sin\phi \, \dot{\theta} - \cos\phi \sin\theta \, \dot{\psi}$$

$$\omega_3 = \dot{\phi} + \cos\theta \dot{\psi} \tag{9.100}$$

(see Problem 9.24).

9.6.6 The Heavy Top

A rigid body rotating about some point O_B under the influence of a torque produced by its weight is called a heavy top. We consider again the special case of a rigid body with axial symmetry, $I_1 = I_2 < I_3$. However, we assume that the fixed point O_B does not coincide with the center of mass, but that it lies on the symmetry axis. The principal moment of inertia I_3 must therefore be evaluated relative to O_B and not relative to the center of mass. The principal moment of inertia I_1 remains unchanged.

If $\mathbf{r}$ is the vector from O_B to the center of mass, the torque on the rigid body is given by

$$\mathbf{N} = \mathbf{r} \times M\mathbf{g} \tag{9.101}$$

The Eulerian angles are now very useful. Inspecting Figure 9.10 we see that the only nonvanishing component of $\mathbf{N}$ is

$$N_\xi = Mgl \sin\theta$$

where $l = |\mathbf{r}|$. There is no torque about the body 3-axis or about any other axis perpendicular to the ξ-axis. It follows that two of the three components of the angular momentum vector are conserved. The components of the torque $\mathbf{N}$ in the body frame S_B are

$$N_1 = N_\xi \cos\psi = Mgl \sin\theta \cos\psi, \tag{9.102}$$

$$N_2 = -N_\xi \sin\psi = -Mgl \sin\theta \sin\psi, \tag{9.103}$$

$$N_3 = 0 \tag{9.104}$$

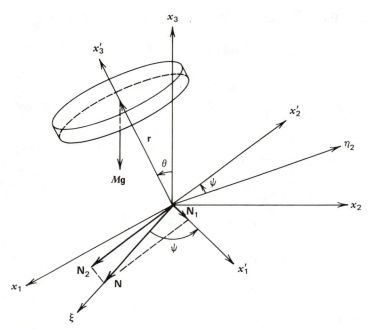

Figure 9.10 The heavy top.

(see Figure 9.10). Substituting Eqs. 9.99 and 9.102 through 9.104 into Euler's equations (9.72) we obtain

$$I_1 \frac{d}{dt}(\sin\theta\sin\psi\dot\phi + \cos\psi\dot\theta) + (I_3 - I_1)(\cos\theta\dot\phi + \dot\psi)(\sin\theta\cos\psi\dot\phi - \sin\psi\dot\theta)$$

$$= Mgl\sin\theta\cos\psi \tag{9.105}$$

$$I_2 \frac{d}{dt}(\sin\theta\cos\psi\dot\phi - \sin\psi\dot\theta) - (I_3 - I_1)(\cos\theta\dot\phi + \dot\psi)(\sin\theta\sin\psi\dot\phi + \cos\psi\dot\theta)$$

$$= -Mgl\sin\theta\sin\psi \tag{9.106}$$

$$I_3 \frac{d}{dt}(\cos\theta\dot\phi + \dot\psi) = 0 \tag{9.107}$$

It is possible to obtain three first integrals from these equations (the values of two conserved components of the angular momentum and the value of the total energy) and to eventually arrive at closed form solutions for all angles (see Problem 9.28). But because Euler's equations for the heavy top are much too cumbersome, we shall pursue an alternate and much simpler route by employing Lagrange's formalism.

Using Eqs. 9.99 we obtain for the rotational kinetic energy the expression

$$T = \tfrac{1}{2}I_1(\omega_1^2 + \omega_2^2) + \tfrac{1}{2}I_3\omega_3^2$$

$$= \tfrac{1}{2}I_1\left[(\sin\theta\sin\psi\dot\theta + \cos\psi\dot\theta)^2 + (\sin\theta\cos\psi\dot\phi - \sin\psi\dot\theta)^2\right]$$

$$+ \tfrac{1}{2}I_3(\cos\theta\dot\phi + \dot\psi)^2$$

$$= \tfrac{1}{2}I_1(\sin^2\theta\dot\phi^2 + \dot\theta^2) + \tfrac{1}{2}I_3(\cos\theta\dot\phi + \dot\psi)^2 \tag{9.108}$$

The potential energy is given by

$$V = Mgl\cos\theta, \tag{9.109}$$

and the Lagrangian $L = T - V$ becomes

$$L(\theta, \dot\phi, \dot\theta, \dot\psi) = \tfrac{1}{2}I_1(\sin^2\theta\dot\phi^2 + \dot\theta^2)$$

$$+ \tfrac{1}{2}I_3(\cos\theta\dot\phi + \dot\psi)^2 - Mgl\cos\theta \tag{9.110}$$

We see that the angles ψ and ϕ are ignorable coordinates such that the conjugate momenta, which are here angular momenta, are constants of the motion:

$$p_\psi = \frac{\partial L}{\partial\dot\psi} = I_3(\cos\theta\dot\phi + \dot\psi) = \text{constant}, \tag{9.111}$$

$$p_\phi = \frac{\partial L}{\partial\dot\phi} = I_1\sin^2\theta\dot\phi + I_3\cos\theta(\cos\theta\dot\phi + \dot\psi)$$

$$= I_1\sin^2\theta\dot\phi + p_\psi\cos\theta = \text{constant} \tag{9.112}$$

Comparing Eqs. 9.107 and 9.111 with each other, we see that p_ψ is the angular momentum about the body 3-axis. The total energy

$$E = T + V$$

$$= \tfrac{1}{2}I_1(\sin^2\theta\dot\phi^2 + \dot\theta^2) + \tfrac{1}{2}I_3(\cos\theta\dot\phi + \dot\psi)^2 + Mgl\cos\theta$$

$$= \tfrac{1}{2}I_1(\sin^2\theta\dot\phi^2 + \dot\theta^2) + \tfrac{1}{2}\frac{p_\psi^2}{I_3} + Mgl\cos\theta \tag{9.113}$$

is another constant of the motion. It consists of two independently conserved parts. Because p_ψ is a constant, a change of the potential energy produces a change of the kinetic energy associated with the 1- and 2-axes only.

Equations 9.111 through 9.113 are sufficient to determine the Eulerian angles as functions of the time. Note that Lagrange's equation for the variable θ,

$$\Lambda_\theta L = I_1\ddot\theta - I_1\sin\theta\cos\theta\dot\phi^2 + I_3(\cos\theta\dot\phi + \dot\psi)\dot\phi\sin\theta - Mgl\sin\theta = 0 \tag{9.114}$$

is not explicitly needed because the energy integral (9.113) is already a consequence of all three Lagrange's equations ((9.114) included). Solving Eq. 9.112 for $\dot{\phi}$, we find

$$\dot{\phi} = \frac{1}{I_1} \frac{(p_\phi - p_\psi \cos\theta)}{\sin^2\theta} \tag{9.115}$$

Elimination of $\dot{\phi}$ from (9.111) results in

$$\dot{\psi} = \frac{p_\psi}{I_3} - \frac{1}{I_1} \frac{(p_\phi - p_\psi \cos\theta)\cos\theta}{\sin^2\theta}, \tag{9.116}$$

whereas elimination of $\dot{\phi}$ from (9.113) leads to

$$\sin^2\theta\dot{\theta}^2 = \left[\frac{2}{I_1}\left(E - \frac{p_\psi^2}{2I_3} \right) - \frac{2Mgl}{I_1}\cos\theta \right]\sin^2\theta$$

$$- \left(\frac{p_\psi}{I_1} \right)^2\left(\frac{p_\phi}{p_\psi} - \cos\theta \right)^2 \tag{9.117}$$

The last equation is a differential equation for θ. If $\theta(t)$ is known, Eqs. 9.115 and 9.116 may then be solved for ϕ and ψ. Unfortunately, the integration of (9.117) involves elliptic integrals, and the extremely complicated calculations are not very informative. We therefore restrict ourselves to a qualitative description of the motion of the heavy top.

Introducing the variable

$$u = \cos\theta \tag{9.118}$$

such that $\dot{u} = -\sin\theta\dot{\theta}$, and defining the constants

$$A = \frac{2I_1Mgl}{p_\psi}, \quad B = \frac{E - \dfrac{p_\psi^2}{2I_3}}{Mgl}, \quad C = \frac{p_\phi}{p_\psi}, \quad D\left(\frac{p_\psi}{I_1} \right)^2, \tag{9.119}$$

we can write (9.117) in the form

$$\dot{u}^2 = D\left[A(B - u)(1 - u^2) - (C - u)^2 \right] \tag{9.120}$$

The term in brackets is a polynomial of degree three in the variable u:

$$f(u) = A(B - u)(1 - u^2) - (C - u)^2 \tag{9.121}$$

Figures 9.11 through 9.15 show the graphs of functions $f(u)$ representing five characteristic motions of the spinning top. The range of physically allowed angles

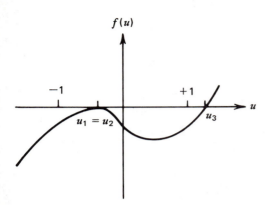

Figure 9.11 Regular precession.

θ is determined by two conditions. First, by definition of the variable u, the interval $[u_1, u_2]$ and the points u_i corresponding to allowed angles must be contained in the interval $[-1, +1]$. Second, because $\dot\theta^2 \geqslant 0$, the function values of $f(u)$ must be positive or zero.

Let us first investigate the case of regular precession which is defined by the condition $\theta = \theta_1 = $ constant $\neq 0$ (and $\neq \pi$). We then have $\dot\theta = 0$ and $\ddot\theta = 0$, and (9.114) becomes

$$\left(I_1\cos\theta_1 \dot\phi^2 - I_3\omega_3\dot\phi + Mgl \right)\sin\theta_1 = 0 \tag{9.122}$$

Solving this equation for $\dot\phi$ we obtain

$$\dot\phi = \frac{I_3\omega_3}{2I_1\cos\theta_1}\left(1 \pm \sqrt{1 - \frac{4I_1 Mgl\cos\theta_1}{I_3^2\omega_3^2}} \right) \tag{9.123}$$

Because $\dot\phi$ is real, the radicand must not be negative. This is always the case for $\pi/2 < \theta_1 < \pi$ (Figure 9.11), and the heavy top processes for all values of ω_3. For

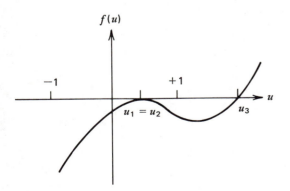

Figure 9.12 Regular precession only if $\omega_3 \geqslant \omega_{\text{crit}}$.

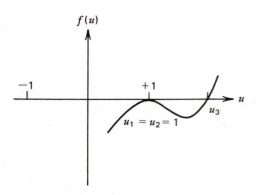

Figure 9.13 The sleeping top.

$0 < \theta_1 < \pi/2$ (Figure 9.12) the radicand is not negative only if

$$\omega_3 \geqslant \frac{2}{I_3}\sqrt{I_1 Mgl \cos \theta_1} = \omega_{\text{critical}},\tag{9.124}$$

and regular precession occurs only if this condition is satisfied. By continuity of the cosine function this result also applies to the case $\theta_1 = 0$ for the "fast vertical" or "sleeping" top (Figure 9.13).

We see from (9.123) that the spinning top may precess with two different angular velocities. If $\omega_3 \gg \omega_{\text{crit}}$, the two limiting values are

$$\dot{\phi}_f = \frac{I_3 \omega_3}{I_1 \cos \theta_1}, \qquad \text{for fast precession}\tag{9.125}$$

and

$$\dot{\phi}_s = \frac{Mgl}{I_1 \omega_3} \qquad \text{for slow precession}\tag{9.126}$$

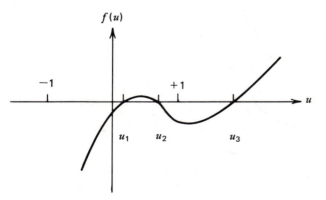

Figure 9.14 Nutation.

Usually, only slow precession is observed because $\dot{\phi}_f$ would become infinity as θ_1 approaches $\pi/2$, whereas $\dot{\phi}_s$ is independent of θ_1. For $\theta_1 = \pi/2$ we obtain from (9.122) only the single value (9.126) such that $\dot{\phi}_s$ is continuous at $\pi/2$.

In the general case (Figure 9.14), the symmetry axis oscillates between the two limiting values θ_1 and θ_2. This oscillation is called nutation, hence the name nutation angle for θ. We conclude from (9.115) that the angular velocity of precession $\dot{\phi}$ is no longer a constant. Depending on the values of p_ϕ and p_ψ and on the interval $[\theta_2, \theta_1]$, there are three types of nutation: $\dot{\phi}$ may be positive at all times, or it may become zero at periodic intervals, or it may vary between positive and negative values. As the result of the combined angular motion, the symmetry axis of the heavy top traces out one of the three possible curves, called the locus of the figure axis, on a sphere as shown in Figures 9.16a, 9.16b, and 9.16c. The curve shown in Figure 9.16b is the one observed if the spinning top is released with $\dot{\theta} = 0$ at the initial angle θ_2.

In the cases depicted in Figures 9.13 and 9.15, the polynomial $f(u)$ has a double root at $u = 1$. This is possible only by (9.121) and (9.119) if simultaneously

$$C = 1, \quad \text{that is,} \ p_\phi = p_\psi,$$

and

$$B = 1, \quad \text{that is,} \quad E = Mgl + \frac{1}{2}\frac{p_\psi^2}{I_3}$$

The polynomial $f(u)$ can then be written as

$$f(u) = [A(1 + u) - 1](1 - u^2),$$

and the third root of $f(u)$ is

$$u = \frac{1}{A} - 1 = \frac{p_\psi^2}{2I_1 Mgl} - 1$$

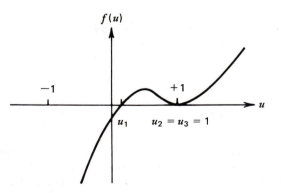

Figure 9.15 Nutation.

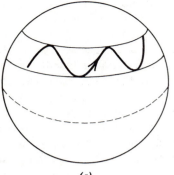

(a)

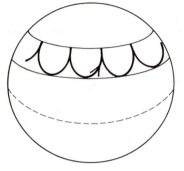

(b)

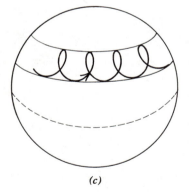

(c)

Figure 9.16 Locus of the figure axis for the three possible types of nutation.

If $u = u_3 > 1$, that is, if

$$\omega_3 > \frac{2}{I_3} \sqrt{I_1 Mgl} \tag{9.127}$$

the corresponding angle θ_3 is unphysical (see Figure 9.13), and the rotation is stable about the vertical axis. We mentioned this case of the sleeping top above following Eq. 9.124. If $u_1 \leqslant u \leqslant 1$, the heavy top performs nutations between the angles $\theta_2 = 0$ and $\theta_1 = \arccos[(p_\psi^2/2I_1 Mgl) - 1]$ (see Figure 9.15). In reality, a sleeping top will not remain in that state because friction causes the angular velocity ω_3 to decrease below the value given by the inequality (9.127), and the top will begin to perform nutations with larger and larger amplitudes.

PROBLEMS

9.1 Show that the number of degrees of freedom is six for a rigid body consisting of three or more mass points. Count the number of independent holonomic conditions as a function of the number N of mass points. Start with $N = 2$.

9.2 Consider two points P and Q on a rigid body that move with angular velocity ω relative to a fixed point O_B on the body. Let ω' be the angular velocity of P relative to Q. Apply the equation for the velocity transformation between two moving reference frames and show that $\omega' = \omega$.

9.3 Consider a rigid body falling freely in a constant uniform gravitational field. Choose an arbitrary point O_B on the body that is not the center of mass. What is the net torque produced by the gravitational force relative to O_B? Compare your answer with the statement proven in Section 9.3 and explain the difference, if any,

9.4 Calculate the moments of inertia for the following objects:

(a) A sphere or radius R about an axis through the center.
(b) A torus with distance a from the axis of symmetry to the centroid, and distance b from the centroid to the surface, about the axis of symmetry. Assume uniform density.

9.5 We have seen that the inertia tensor may be diagonalized by rotating the coordinate system about its origin O_B. It is sometimes possible to obtain an inertia tensor in diagonal form by a translation of the coordinate system from O_{CM} to some point O, whereby the directions of the coordinate axes remain unchanged.

(a) Apply Eq. 9.37 to derive a condition for the vanishing of the off-diagonal elements of I in terms of the elements I'_{kl}. Find the coordinates of the origin of the new coordinate system. There are two solutions. Also find I_{ll} in terms of the elements I'_{kl}.
(b) Consider the following mass distribution in S_{CM}. $m_i = 1/2$, $i = 1, 2, \ldots$ 6; $r'_1 = (1, -1, 0)$, $r'_2 = (-1, 1, 0)$, $r'_3 = (1, 0, -1)$, $r'_4 = (-1, 0, 1)$, $r'_5 = (0, 1, -1)$, $r'_6 = (0, -1, 1)$. Find the coordinates of the origins of the systems relative to which $I_{kl} = 0$ for $k \neq 1$. Also calculate the inertia tensor in these coordinate systems.

9.6 Consider the system described in Examples 9.1 and 9.6. Graph the mass distribution and a straight line making an angle of $\theta = \arctan(1/2) = 26.6°$ with respect to the horizontal. Do you expect this line to be the axis about which the moment of inertia of the system assumes a minimum value? Now, using calculus, and performing a rotation about the x_2-axis, determine the

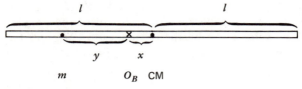

Figure 9.17 Problem 9.12.

exact angle of rotation such that the moment of inertia about the new x_1-axis assumes a minimum value.

9.7 Evaluate Eq. 9.41 for the system described in Examples 9.1 and 9.7.

9.8 Derive Eq. 9.46.

9.9 Calculate the moment of inertia of a cylinder of radius R, height h, and density $\rho(r) = kr$, where k is a constant about the symmetry axis. Express the answer in terms of R, h, and the mass M of the cylinder.

9.10 Find the dimensions of a circular cylinder such that the inertia ellipsoid becomes a sphere.

9.11 (a) A spherical shell (mass m, inner radius r_1, outer radius r_2) rolls down an inclined plane without slipping. Find the value of the acceleration along the inclined plane.
(b) Find the acceleration of a cylindrical shell under the same conditions.

9.12 A uniform rigid rod of length $2l$ and mass M is free to rotate about an axis perpendicular to the rod through the point O_B, a distance x from the center of mass. The rod is to be balanced by an additional mass m positioned in such a way that O_B becomes the center of mass, and that the moments of inertia of both sides of the rod relative to O_B become equal to each other (see Figure 9.17). Find the value of m and y.

9.13 Derive Eqs. 9.59 and 9.60.

9.14 Find the maximum frequency of oscillation of a compound pendulum in terms of g and the radius of gyration k_{CM}.

9.15 A circular hoop of radius R is allowed to rotate about an axis passing through two points on the circumference (see Figure 9.18). If the axis is

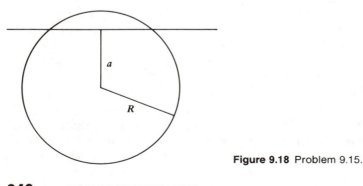

Figure 9.18 Problem 9.15.

horizontal and the hoop swings as a physical pendulum, find the period of oscillation for small amplitudes.

9.16 Where must a billiard ball be struck horizontally such that it starts rolling without slipping? Repeat the problem for a cylinder.

9.17 The constant values for T_{rot}, $\mathbf{L}$, ω^2, and ω indicated in Table 9.2 are not all independent of each other. State all identities between these constants for the three cases $I_1 < I_2 < I_3$, $I_1 = I_2 < I_3$, and $I_1 = I_2 = I_3$.

9.18 (a) Let $T_{kl}(t)$ be the components of a time dependent tensor of rank two. Derive the equation

$$\left(\frac{dT_{kl}}{dt}\right)_S = \left(\frac{dT_{kl}}{dt}\right)_B + \sum_{m=1}^{3}\left(\Omega_{km}T_{ml} - T_{km}\Omega_{ml}\right)$$

which reads in matrix form

$$\left(\frac{d\mathbf{T}}{dt}\right)_S = \left(\frac{d\mathbf{T}}{dt}\right)_B + \Omega\mathbf{T} - \mathbf{T}\Omega$$

The algebraic structure of this equation is typical for matrices and operators in many areas of physics (see Appendix 4). *Hint*: Apply (2.89) and look up the derivation of (3.31).

(b) Let T_{kl} be a tensor in diagonal form: $T_{kl} = d_k\delta_{kl}$. Show that

$$\sum_{l=1}^{3}\left(\frac{dT_{1l}}{dt}\right)\omega_l = \omega_2\omega_3(d_3 - d_2)$$

(c) Derive Euler's equations from

$$\frac{d}{dt}\left(\sum_{l=1}^{3} I_{kl}\omega_l\right) = N_k$$

using the result of part (b).

9.19 Using Euler's equations, show that the torques of constraint vanish if the rigid body rotates about a principal axis.

9.20 Derive the rotational work-energy theorem from Euler's equations.

9.21 Show that a rigid body with $I_1 = I_2 < I_3$ or $I_1 < I_2 = I_3$ possesses only one stable axis of (torque-free) rotation.

9.22 For the rigid body described in Section 6.4 find $|\omega|$ in terms of L, T, I_1, and I_3.

9.23 For the torque-free rigid body with $I_1 = I_2 < I_3$ find the angles between ω and $\mathbf{L}$, and $\omega_3\mathbf{e}_3$ and $\mathbf{L}$ in terms of I_1, I_3, ω_0, and ω_3.

9.24 (a) Prove the equation $\mathbf{R}_\phi^T\Omega_\phi\mathbf{R}_\phi = \Omega_\phi$. What does it mean in the vector formalism?

(b) Derive Eqs. 9.100. This may be done, for example, by evaluating $\Omega = \mathbf{R}^T\Omega'\mathbf{R}$, where $\mathbf{R}$ is given by (9.94). Remember that $\mathbf{A}\mathbf{A}^T = \mathbf{A}^T\mathbf{A} = 1$ for an orthogonal matrix.

9.25 Write Eq. 9.98 in the vector formalism. Show that the components of the vector $\boldsymbol{\omega}$ in S_B are those of (9.99).

9.26 In order to test your perception of rotation, perform the following experiment. Construct a Cartesian coordinate system out of wire and label the axes x, y, and z. Perform three finite rotations in succession using the Eulerian angles $\phi = 45°$, $\theta = 45°$, and $\psi = 45°$. Compare the initial position with the final position. The combined rotations can be regarded as one rotation about a fixed axis defined by the radius vector $\mathbf{r} = (1, 0, 1 + \sqrt{2}/2)$ with the angle of rotation γ. From your experiment, would you expect γ to be smaller than, equal to, or larger than $90°$? Trace the orbit of the endpoint of the x-axis of your model. Compare your guessed answer with the exact value obtained from the formula

$$\cos \gamma = \tfrac{1}{2}(\cos \theta + 1)(\cos(\phi + \psi) + 1) - 1$$

Can you explain the discrepancy, if any?

9.27 A symmetrical top with $I_1 = I_2 < I_3$ is acted upon by a torque $\mathbf{N} = N_3(t)\mathbf{e}_3'$. Develop a procedure to solve Euler's equations.

9.28 (a) Derive the two equations

$$I_1(\ddot{\theta} - \sin \theta \cos \theta \dot{\phi}^2) + I_3(\cos \theta \dot{\phi} + \dot{\psi})\sin \theta \dot{\phi} - Mgl \sin \theta = 0 \quad (A)$$

and

$$I_1(\sin \theta \ddot{\phi} + 2 \cos \theta \dot{\phi} \ \dot{\theta}) - I_3(\cos \theta \dot{\phi} + \dot{\psi})\dot{\theta} = 0 \qquad (B)$$

from (9.105) and (9.106).
(b) Derive Eq. 9.112 from (B).
(c) Derive Eq. 9.113 from (A) and (B). Note that $p_\psi^2/2I_3 = $ constant.

9.29 Set up Hamilton's equations for the heavy top in terms of the Eulerian angles and the conjugate momenta.

chapter
10 Coupled Oscillators

Beauty is in the eye of the beholder.

MARGARET WOLFE HUNGERFORD

10.1 INTRODUCTION

Many systems consist of particles oscillating about their points of equilibrium under the influence of forces between these particles. The simplest such systems involve two particles, for example, two masses connected by a spring (Figure 10.1), the double pendulum (Figure 10.2), or two coupled pendulums (Figure 10.3). A far more complicated system is the real solid. It was shown in Chapter 9 that a real solid could be treated as a rigid body under certain simplifying conditions, but that a rigid body does not actually exist. A more realistic model of a solid is one in which the atoms are considered as particles subjected to elastic

1 ⎯⎯⎯〰〰〰〰〰⎯⎯⎯ 2

Figure 10.1 Two coupled mass points.

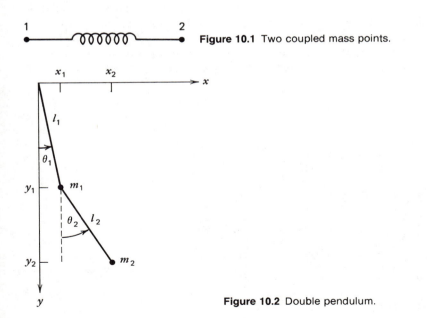

Figure 10.2 Double pendulum.

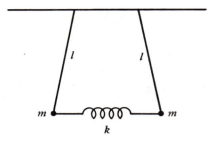

Figure 10.3 Coupled pendulum.

forces between the atoms and oscillating about their points of equilibrium. A simplified model of a cubic crystal is shown in Figure 10.4.

In this chapter we shall investigate the possible modes and frequencies of oscillation of coupled systems. The information obtained may also be useful in other areas of physics, for example, for the calculation of the specific heats of solids. We shall see that Lagrange's formalism is a very powerful and convenient tool for treating coupled oscillators, whether the number of particles is small or large.

10.2 THE EQUATIONS OF MOTION

We consider a system of several interacting particles described by a set $\{q_1, \ldots q_f\}$ of generalized coordinates. We assume that there are no frictional forces and that the forces between the particles are conservative. The potential energy function associated with a pair of particles is, in general, not of the harmonic oscillator type $(1/2)kx^2$. In order to obtain a set of manageable equations of motion for the system, we require that each particle performs only small oscillations about its equilibrium position. All we need to do now is in the Lagrangian formalism to generalize the procedure developed in Section 5.2 for small oscillations of one mass point to many particle systems.

Let $V(q_1, \ldots q_f)$ be the potential energy function of the system. By (8.20) and (8.21), the system is in equilibrium if all generalized forces Q_j vanish:

$$Q_j = -\frac{\partial V}{\partial q_j} = 0, \quad j = 1, \cdots f \tag{10.1}$$

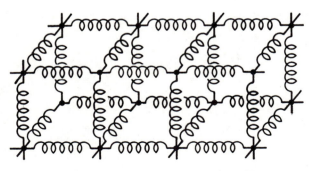

Figure 10.4 Simplified model of a cubic crystal.

These equations allow us to calculate the coordinates q_{0_j} of the equilibrium position of the system. The system is in stable equilibrium if V assumes a minimum at $q_{0_j}, j = 1, \ldots f$.

Expanding the potential energy function in a Taylor series about the equilibrium position, we find

$$V(q_j) = V(q_{0_j}) + \frac{1}{1!} \sum_{k=1}^{f} \frac{\partial V}{\partial q_k}\bigg|_{q_k = q_{0_k}} (q_k - q_{0_k})$$

$$+ \frac{1}{2!} \sum_{k=1}^{f} \sum_{l=1}^{f} \frac{\partial^2 V}{\partial q_k \, \partial q_l}\bigg|_{\substack{q_k = q_{l_k} \\ q_l = q_{0_l}}} (q_k - q_{0_k})(q_l - q_{0_l})$$

$$+ \text{terms of higher order.} \tag{10.2}$$

The first term $V(q_{0_j})$ is a constant that may be set equal to zero because it is of no consequence for the equation of motion. The second term vanishes because of (10.1). All terms of higher order are negligible under the assumption that the derivatives of V at q_{0_j} are at most of the same order of magnitude as the second derivatives. The constants

$$V_{kl} = \frac{\partial^2 V}{\partial q_k \, \partial q_l}\bigg|_{\substack{q_k = q_{0k} \\ q_l = q_{0_l}}} = V_{lk} \tag{10.3}$$

form a symmetric matrix $\mathbf{V}$ and correspond to the spring constant k in Eq. 6.15. Choosing new generalized coordinates

$$\xi_k = q_k - q_{0_k}, \qquad k = 1, \ldots f, \tag{10.4}$$

such that the equilibrium position is located at $\xi_k = 0$, we obtain from (10.2), (10.3), and (10.4)

$$V(\xi_k) = \frac{1}{2} \sum_{k=1}^{f} \sum_{l=1}^{f} V_{kl} \xi_k \xi_l \tag{10.5}$$

Because in the neighborhood of the potential energy minimum (the point of stable equilibrium) $V(\xi_1, \ldots \xi_f) > V(0, \ldots 0)$ holds for any choice of the ξ_k, not all being equal to zero, the homogeneous quadratic form (10.5) is positive definite. The necessary and sufficient conditions that a homogeneous quadratic form be positive definite are that the coefficients $V_{kl} = V_{lk}$ satisfy[1]

$$V_{11} > 0, \quad \begin{vmatrix} V_{11} & V_{12} \\ V_{21} & V_{22} \end{vmatrix} > 0, \quad \begin{vmatrix} V_{11} & V_{12} & V_{13} \\ V_{21} & V_{22} & V_{23} \\ V_{31} & V_{32} & V_{33} \end{vmatrix} > 0, \text{ and so on.} \tag{10.6}$$

(The V_{kl} need not all be positive, however.)

---

[1] A simple proof for the case $f = 3$ is given in D. Widder, *Advanced Calculus*, Prentice-Hall, Englewood Cliffs, N.J., 1961.

Example 10.1

Find the components V_{kl} of V for the double pendulum (Figure 10.2).

In terms of the generalized coordinates θ_1 and θ_2, the potential energy is given by

$$V = -m_1 g y_1 - m_2 g y_2$$

$$= -m_1 g l_1 \cos \theta_1 - m_2 g (l_1 \cos \theta_1 + l_2 \cos \theta_2)$$

Evaluating $\partial V / \partial \theta_1$ and $\partial V / \partial \theta_2$, or simply inspecting Figure 10.2, we obtain for the equilibrium configuration the coordinates

$$\theta_{0_1} = 0 \quad \text{and} \quad \theta_{0_2} = 0$$

The components V_{kl} are

$$V_{11} = \left. \frac{\partial^2 V}{\partial \theta_1^2} \right|_{\substack{\theta_1 = 0 \\ \theta_2 = 0}} = m_1 g l_1 + m_2 g l_1 = (m_1 + m_2) g l_1,$$

$$V_{22} = \left. \frac{\partial^2 V}{\partial \theta_2^2} \right|_{\substack{\theta_1 = 0 \\ \theta_2 = 0}} = m_2 g l_2,$$

$$V_{12} = V_{21} = \frac{\partial^2 V}{\partial \theta_1 \, \partial \theta_2} = 0,$$

such that

$$V = \begin{pmatrix} (m_1 + m_2) g l_1 & 0 \\ 0 & m_2 g l_2 \end{pmatrix} \tag{10.7}$$

Because this matrix satisfies the conditions (10.6), the associated homogeneous quadratic form is positive definite. ∎

If an oscillating system is free to move through space, its motion can be regarded as a superposition of the motion of the center of mass relative to a fixed reference frame, and the motion of the system relative to the center of mass. In the absence of external forces, the center of mass velocity is constant, that is, the center of mass does not undergo oscillations. The form (10.5) is then no longer positive definite but only positive semidefinite. (A homogeneous quadratic form is called positive semidefinite if

$$\sum_{k=1}^{f} \sum_{l=1}^{f} V_{kl} \xi_k \xi_l \geq 0 \quad \text{for all} \quad \xi_k, \qquad k = 1, \ldots f)$$

The form (10.5) is, however, positive definite if it is restricted to the generalized coordinates relative to the center of mass.

Example 10.2

Find the components V_{kl} of **V** for the linear triatomic molecule shown in Figure 10.5.

We consider displacements in the direction of the x-axis only. Each spring is stretched (or compressed) by an amount equal to the relative displacement of the attached masses from their equilibrium positions. Therefore,

$$V = \tfrac{1}{2}k(\xi_2 - \xi_1)^2 + \tfrac{1}{2}k(\xi_3 - \xi_2)^2$$

$$= \tfrac{1}{2}k(\xi_1^2 + 2\xi_2^2 + \xi_3^2 - 2\xi_1\xi_2 - 2\xi_2\xi_3)$$

V is already homogeneous quadratic, hence,

$$\mathbf{V} = k\begin{pmatrix} 1 & -1 & 0 \\ -1 & 2 & -1 \\ 0 & -1 & 1 \end{pmatrix}$$

We have

$$V_{11} = k > 0, \qquad \begin{vmatrix} V_{11} & V_{12} \\ V_{21} & V_{22} \end{vmatrix} = k^2 \begin{vmatrix} 1 & -1 \\ -1 & 2 \end{vmatrix} = k^2 > 0,$$

but

$$\begin{vmatrix} V_{11} & V_{12} & V_{13} \\ V_{21} & V_{22} & V_{23} \\ V_{31} & V_{32} & V_{33} \end{vmatrix} = k^3 \begin{vmatrix} 1 & -1 & 0 \\ -1 & 2 & -1 \\ 0 & -1 & 1 \end{vmatrix} = 0,$$

and the associated form (10.5) is not positive definite. This can also directly be seen from $V = 0$ if $\xi_1 = \xi_2 = \xi_3 = 1$. ∎

Let us now turn to the kinetic energy. In the absence of rheonomous conditions, the kinetic energy T is a homogeneous quadratic form in the generalized velocities $\dot{q}_k$,

$$T = \frac{1}{2} \sum_{k=1}^{f} \sum_{l=1}^{f} a_{kl}(q_j)\dot{q}_k\dot{q}_l, \tag{10.8}$$

where the coefficients a_{kl} are defined by Eq. 8.23. Expanding a_{kl} in a Taylor series about the equilibrium position we obtain

$$a_{kl}(q_j) = a_{kl}(q_{0_j}) + \sum_{m=1}^{f} \left.\frac{\partial a_{kl}}{\partial q_m}\right|_{q_m = q_{0_m}} (q_m - q_{0_m}) + \cdots$$

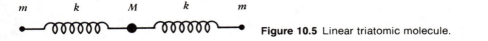

m k M k m

Figure 10.5 Linear triatomic molecule.

Because the kinetic energy (10.8) is already small of second order (it contains the products $\dot{q}_k \dot{q}_l$), the $a_{kl}(q_j)$ may be replaced by $a_{kl}(q_{0_j})$. Defining the components T_{kl} of a symmetric matrix T by

$$T_{kl} = a_{kl}(q_{0_j}) = T_{lk} \tag{10.9}$$

we can write (10.8) in the form

$$T = \frac{1}{2} \sum_{k=1}^{f} \sum_{l=1}^{f} T_{kl} \dot{\xi}_k \dot{\xi}_l \tag{10.10}$$

because $\dot{q}_k = \dot{\xi}_k$. The kinetic energy is, by definition, always positive unless all velocities vanish. We therefore have $T(\dot{\xi}_1, \ldots \dot{\xi}_f) > T(0, \ldots 0) = 0$ for any choice of the $\dot{\xi}_k$, not all being equal to zero, and the homogeneous form (10.10) is always positive definite.

If the kinetic energy for a given system has been obtained in the form (10.10), the component T_{kk} can easily be recognized as the coefficient of $(1/2)\dot{\xi}_k^2$. However, for $k \neq l$, the component T_{kl} is the coefficient of $\dot{\xi}_k \dot{\xi}_l$ because of the symmetry property $T_{kl} = T_{lk}$.

Example 10.3

Find the components T_{kl} of T for the double pendulum (Figure 10.2).

The kinetic energy is

$$\begin{aligned}
T &= \tfrac{1}{2}m_1\left(\dot{x}_1^2 + \dot{y}_1^2\right) + \tfrac{1}{2}m_2\left(\dot{x}_2^2 + \dot{y}_2^2\right) \\
&= \tfrac{1}{2}m_1\left[\left(l_1(-\sin\theta_1)\dot{\theta}_1\right)^2 + \left(l_1\cos\theta_1\dot{\theta}_1\right)^2\right] \\
&\quad + \tfrac{1}{2}m_2\left[\left(l_1(-\sin\theta_1)\dot{\theta}_1 + l_2(-\sin\theta_2)\dot{\theta}_2\right)^2 + \left(l_1\cos\theta_1\dot{\theta}_1 + l_2\cos\theta_2\dot{\theta}_2\right)^2\right] \\
&= \tfrac{1}{2}m_1 l_1^2\dot{\theta}_1^2 + \tfrac{1}{2}m_2\left[l_1^2\dot{\theta}_1^2 + l_2^2\dot{\theta}_2^2 + 2l_1 l_2\cos(\theta_1 - \theta_2)\dot{\theta}_1\dot{\theta}_2\right]
\end{aligned}$$

At the equilibrium point, T reduces to

$$T = \tfrac{1}{2}(m_1 + m_2)l_1^2\dot{\theta}_1^2 + \tfrac{1}{2}m_2 l_2^2\dot{\theta}_2^2 + m_2 l_1 l_2\dot{\theta}_1\dot{\theta}_2$$

The components T_{kk} and T_{kl} are the coefficients of $(1/2)\dot{\theta}_k^2$ and $\dot{\theta}_k\dot{\theta}_l$, respectively:

$$T_{11} = (m_1 + m_2)l_1^2, \qquad T_{22} = m_2 l_2^2, \qquad T_{12} = T_{21} = m_2 l_1 l_2$$

Hence, T is given by

$$T = \begin{pmatrix} (m_1 + m_2)l_1^2 & m_2 l_1 l_2 \\ m_2 l_1 l_2 & m_2 l_2^2 \end{pmatrix} \tag{10.11}$$

∎

Using Eqs. (10.5) and (10.10) we obtain the Lagrangian L in the form

$$L = T - V = \sum_{k=1}^{f} \sum_{l=1}^{f} \frac{1}{2} \left(T_{kl} \dot{\xi}_k \dot{\xi}_l - V_{kl} \xi_k \xi_l \right) \qquad (10.12)$$

from which follow Lagrange's equations for the oscillating system:

$$\Lambda_k L = \frac{d}{dt} \frac{\partial}{\partial \dot{\xi}_k} \sum_{j=1}^{f} \sum_{l=1}^{f} \frac{1}{2} T_{jl} \dot{\xi}_j \dot{\xi}_l - \frac{\partial}{\partial \xi_k} \left(- \sum_{j=1}^{f} \sum_{l=1}^{f} \frac{1}{2} V_{jl} \xi_j \xi_l \right) = 0$$

or

$$\sum_{l=1}^{f} \left(T_{kl} \ddot{\xi}_l + V_{kl} \xi_l \right) = 0, \quad k = 1, \ldots f \qquad (10.13)$$

Equations (10.13) represent a system of f coupled second order linear differential equations with constant coefficients. The general solution of each $\xi_l(t)$ is a linear combination of $2f$ linearly independent solutions of the system and contains $2f$ arbitrary integration constants.

10.3 NORMAL FREQUENCIES

In order to solve Lagrange's equations (10.13) we choose trial functions of the form

$$\xi_l = Ca_l \cos(\omega t + \delta) \qquad (10.14)$$

for each coordinate ξ_l, $l = 1, \ldots f$. The constant C is a scale factor whose meaning becomes clear later. The form (10.14) for ξ_l was chosen because it reduces to the known general solution $\xi = C\cos(\omega t + \delta)$ for the simple harmonic oscillator.

Differentiating ξ_l twice, we obtain

$$\ddot{\xi}_l = -\omega^2 C a_l \cos(\omega t + \delta) \qquad (10.15)$$

Substitution of (10.14) and (10.15) into (10.13) yields the set of f homogeneous linear algebraic equations for the f unknowns a_l:

$$\sum_{l=1}^{f} \left(V_{kl} - \omega^2 T_{kl} \right) a_l = 0, \quad k = 1, \ldots f \qquad (10.16)$$

Two conclusions can be drawn from the fact that the system of equations (10.16) is linear homogeneous. First, if $\{a_l\}$ is a nontrivial solution (i.e., at least one of the a_l is different from zero), then also $\{Ca_l\}$ is a solution, or, in other words, only the ratios $a_l : a_m$ are uniquely determined. Second, it is shown in linear algebra that nontrivial solutions $\{a_l\}$ of (10.16) exist only if

$$\det |V - \omega^2 T| = 0 \qquad (10.17)$$

This equation, called characteristic equation or secular equation, is a polynomial of degree f in ω^2, which has f roots ω_j^2. The ω_j are called normal frequencies of the system. Because damping is absent, that is, energy is conserved, we expect the roots ω_j^2 not to be negative. The normal frequencies ω_j are thus real and can be chosen as $\omega_j \geqslant 0$. They may then be ordered so that

$$0 \leqslant \omega_1 \leqslant \omega_2 \leqslant \ldots \leqslant \omega_f$$

If the ω_j are all different, the frequency spectrum is called nondegenerate. If some (or all) of the ω_j are equal, the frequency spectrum is called degenerate.

The fact that the ω_j are real can be shown as follows. If we multiply Eq. 10.16 by a_k and extend the summation over k, we obtain

$$\sum_{k=1}^{f} \sum_{l=1}^{f} \left(V_{kl} - \omega^2 T_{kl} \right) a_k a_l = 0,$$

such that

$$\omega^2 = \frac{\displaystyle\sum_{k=1}^{f} \sum_{l=1}^{f} V_{kl} a_k a_l}{\displaystyle\sum_{k=1}^{f} \sum_{l=1}^{f} T_{kl} a_k a_l} \qquad (10.18)$$

All terms on the right side of (10.18) are real, hence ω^2 must be real also. Because the elements T_{kl} define a positive definite form, the denominator is always positive unless all a_k are equal to zero. The latter case is of no interest because all particles would be at rest. Note that the denominator is not twice the kinetic energy. If the coefficients V_{kl} define a positive definite form, the numerator of (10.18) is always positive for at least one $a_k \neq 0$, which implies that ω^2 is positive. If the coefficients V_{kl} define a positive semidefinite form, the numerator may become zero for some $a_k \neq 0$ in which case at least one frequency ω_j vanishes. Then there exist some generalized coordinates for which the associated generalized velocities remain constant.

Example 10.4

Find the normal frequencies of the double pendulum.
Inserting the results of Examples 10.1 and 10.3 into Eq. 10.17 we have

$$\begin{vmatrix} (m_1 + m_2)gl_1 - \omega^2(m_1 + m_2)l_1^2 & -\omega^2 m_2 l_1 l_2 \\ -\omega^2 m_2 l_1 l_2 & m_2 g l_2 - \omega^2 m_2 l_2^2 \end{vmatrix} = 0,$$

which simplifies to

$$m_2^2 l_1^2 l_2^2 \begin{vmatrix} \left(\dfrac{m_1}{m_2} + 1\right)\dfrac{g}{l_1} - \left(\dfrac{m_1}{m_2} + 1\right)\omega^2 & -\omega^2 \\[3mm] -\omega^2 & \dfrac{g}{l_2} - \omega^2 \end{vmatrix} = 0$$

by the factor rules of determinants. The determinantal equation leads to

$$\omega^4 - \omega^2\left(1 + \frac{m_2}{m_1}\right)\left(\frac{1}{l_1} + \frac{1}{l_2}\right)g + \left(1 + \frac{m_2}{m_1}\right)\frac{g^2}{l_1 l_2} = 0$$

which has the roots

$$\omega_{1,2}^2 = \frac{1}{2}\left(1 + \frac{m_2}{m_1}\right)\left(\frac{1}{l_1} + \frac{1}{l_2}\right)g\left[1 \pm \left(1 - 4\frac{l_1 l_2}{(l_1 + l_2)^2}\frac{m_1}{(m_1 + m_2)}\right)^{1/2}\right]$$

Note that $\omega_1 \neq \omega_2$ for any value of l_1, l_2, m_1, and m_2. If, in particular, $l_1 = l_2 = l$ and $m_1 = m_2 = m$, we obtain

$$\omega_1^2 = \frac{g}{l}(2 - \sqrt{2}) \quad \text{and} \quad \omega_2^2 = \frac{g}{l}(2 + \sqrt{2}) \tag{10.19}$$

∎

10.4 THE GENERAL SOLUTION OF THE EQUATIONS OF MOTION

In this section, we shall derive the general solution of the equations of motion (10.13) in an elementary although somewhat cumbersome way. The effort spent will not be wasted, however, because the insights gained here will help us to better understand the algebraic properties of the equations of motion and possible shortcuts for their solution. We shall discuss these topics in Section 10.5.

Let us consider the case of a nondegenerate frequency spectrum, that is, the normal frequencies are all different.

For each frequency ω_j there exists a solution of the form (10.14) with scale factor C_j, coefficients a_{lj} and phase angle δ_j. The set $\{a_{lj}\}$ is called the normal mode associated with the frequency ω_j, or the jth normal mode, for short. Because Eq. 10.14 is already a linear combination of the two linearly independent functions $\cos \omega t$ and $\sin \omega t$, the most general solution ξ_l of (10.13), which must consist of $2f$ linearly independent functions, is of the form

$$\xi_l(t) = \sum_{j=1}^{f} a_{lj} C_j \cos(\omega_j t + \delta_j), \qquad l = 1, \ldots f, \tag{10.20}$$

where the sum is taken over the index j corresponding to the f frequencies ω_j. The product $a_{lj}C_j$ is the amplitude of that part of ξ_l which oscillates with the frequency ω_j. Equation 10.20 may also be written

$$\xi_l(t) = \sum_{j=1}^{f} a_{lj}\left(D_j\cos \omega_j t + E_j\sin \omega_j t\right) \qquad (10.21)$$

where

$$D_j = C_j\cos \delta_j \qquad (10.22)$$

and

$$E_j = -C_j\sin \delta_j \qquad (10.23)$$

All constants appearing in Eq. 10.20 or 10.21 can now be determined by the following operations performed in the indicated order.

Step 1. Calculate the normal frequencies ω_j from the characteristic equation (10.17).
Step 2. Replace ω^2 in (10.16) by ω_j^2, $j = 1, \ldots f$, and calculate the f sets of solutions $\{a_{lj}\}$, one for each j. The a_{lj} contain a common factor that may be fixed by arbitrarily setting $a_{1j} = 1$ if $a_{1j} \neq 0$. If $a_{1j} = 0$ but $a_{2j} \neq 0$, one may set $a_{2j} = 1$, and so on. (Another procedure for fixing the a_{lj} is presented in the next section.)
Step 3. Calculate D_j and E_j from the initial conditions

$$\xi_l(0) \equiv \xi_{l_0} = \sum_{j=1}^{f} a_{lj}D_j \qquad (10.24)$$

and

$$\dot{\xi}_l(0) \equiv v_{l_0} = \sum_{j=1}^{f} a_{lj}\omega_j E_j \qquad (10.25)$$

imposed on the system. Equations 10.24 and 10.25 each represent sets of f inhomogeneous linear equations for the f unknowns D_j and E_j, respectively. In the next section we shall derive equations that yield D_j and E_j explicitly.

The task of determining the constants D_j and E_j (or C_j and δ_j) would be greatly simplified, especially if the number of degrees of freedom is very large, if we had simply

$$a_{lj} = \delta_{lj}$$

Equations 10.21 through 10.23 would then reduce to

$$\xi_l(t) = D_l \cos \omega_l t + E_l \sin \omega_l t, \qquad (10.26)$$

$$\xi_l(0) \equiv \xi_{l_0} = D_l \qquad (10.27)$$

and

$$\dot{\xi}_l(0) \equiv v_{l_0} = \omega_l E_l, \qquad (10.28)$$

respectively. Coordinates for which (10.26) holds are called normal coordinates. It turns out that it is possible to find a coordinate transformation such that *all* ξ_l are of the form (10.26). The reason why this can be done is hidden in the algebraic structure of Eqs. 10.16. We shall discuss the properties of normal coordinates in Section 10.6.

Example 10.5

Find the normal modes for the double pendulum if $l_1 = l_2 = l$ and $m_1 = m_2 = m$.

Using the results of Examples 10.1 and 10.3, and substituting ω_1^2, Eq. 10.19, for ω^2 into (10.16), we find

$$\begin{pmatrix} 2mgl - \frac{g}{l}(2 - \sqrt{2})2ml^2 & -\frac{g}{l}(2 - \sqrt{2})ml^2 \\ -\frac{g}{l}(2 - \sqrt{2})ml^2 & mgl - \frac{g}{l}(2 - \sqrt{2})ml^2 \end{pmatrix} \begin{pmatrix} a_{11} \\ a_{21} \end{pmatrix} = 0$$

which is equivalent to

$$(2 - 2\sqrt{2})a_{11} + (2 - \sqrt{2})a_{21} = 0$$

and

$$(2 - \sqrt{2})a_{11} + (1 - \sqrt{2})a_{21} = 0$$

A possible solution is

$$a_{11} = 1, \qquad a_{21} = \sqrt{2} \qquad (10.29)$$

Similarly, for ω_2^2 we obtain

$$a_{12} = 1, \qquad a_{22} = -\sqrt{2} \qquad (10.30)$$

Because a_{11} and a_{12} relate to particle 1, whereas a_{21} and a_{22} relate to particle 2, the results (10.29) and (10.30) imply that the particles move simultaneously in the same direction with the smaller frequency ω_1 and in the opposite direction with the larger frequency ω_2 (see Figs. 10.6a and 10.6b). ∎

Example 10.6

Find the general solution of the equations of motion for the double pendulum described in Example 10.5 if the initial values are:

$$\xi_1(0) = \xi_{1_0}, \quad \xi_2(0) = \xi_{2_0}, \quad \dot{\xi}_1(0) = \nu_{1_0}, \quad \dot{\xi}_2(0) = \nu_{2_0}$$

Remember that the generalized coordinates ξ_l represent angles, whereas the $\dot{\xi}_l$ are angular velocities.

Equations 10.24 applied to our problem read

$$a_{11} D_1 + a_{12} D_2 = D_1 + D_2 = \xi_{1_0}$$

$$a_{21} D_1 + a_{22} D_2 = \sqrt{2}\, D_1 - \sqrt{2}\, D_2 = \xi_{2_0}$$

The solutions are

$$D_1 = \frac{1}{2}\left(\xi_{1_0} + \frac{1}{\sqrt{2}}\xi_{2_0}\right) \quad \text{and} \quad D_2 = \frac{1}{2}\left(\xi_{1_0} - \frac{1}{\sqrt{2}}\xi_{2_0}\right)$$

From (10.25) we obtain

$$E_1 = \frac{1}{2\omega_1}\left(\nu_{1_0} + \frac{1}{\sqrt{2}}\nu_{2_0}\right) \quad \text{and} \quad E_2 = \frac{1}{2\omega_2}\left(\nu_{1_0} - \frac{1}{\sqrt{2}}\nu_{2_0}\right),$$

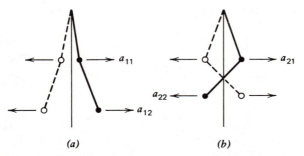

(a) (b)

Figure 10.6 (a) Double pendulum — normal mode corresponding to ω_1. (b) Double pendulum — normal mode corresponding to ω_2.

254 COUPLED OSCILLATORS

and the general solution (10.21) is therefore

$$\xi_1(t) = \frac{1}{2}\left\{\left(\xi_{1_0} + \frac{1}{\sqrt{2}}\xi_{2_0}\right)\cos \omega_1 t + \left(\xi_{1_0} - \frac{1}{\sqrt{2}}\xi_{2_0}\right)\cos \omega_2 t\right.$$

$$\left. + \frac{1}{\omega_1}\left(\nu_{1_0} + \frac{1}{\sqrt{2}}\nu_{2_0}\right)\sin \omega_1 t + \frac{1}{\omega_2}\left(\nu_{1_0} - \frac{1}{\sqrt{2}}\nu_{2_0}\right)\sin \omega_2 t\right\}$$

$$\xi_2(t) = \frac{1}{2}\left\{\left(\sqrt{2}\,\xi_{1_0} + \xi_{2_0}\right)\cos \omega_1 t - \left(\sqrt{2}\,\xi_{1_0} - \xi_{2_0}\right)\cos \omega_2 t\right.$$

$$\left. + \frac{1}{\omega_1}\left(\sqrt{2}\,\nu_{1_0} + \nu_{2_0}\right)\sin \omega_1 t - \frac{1}{\omega_2}\left(\sqrt{2}\,\nu_{1_0} - \nu_{2_0}\right)\sin \omega_2 t\right\} \qquad \blacksquare$$

10.5 ALGEBRAIC PROPERTIES OF THE EQUATIONS OF MOTION

Let us take a closer look at the system of Eqs. 10.16. We recognize that these equations define an eigenvalue problem with ω^2 as the eigenvalue. Obviously, because the unit matrix has been replaced by the matrix T, the eigenvalue problem (10.16) is not posed in the standard form (2.91).

We shall now outline a general procedure which enables us, first, to transform (10.16) into the standard form, second, to present the general solution of (10.16) explicitly, and, third, to obtain the complete set of normal coordinates for the system. Our objective is equivalent to proving the following mathematical theorem: Two symmetric matrices can simultaneously be diagonalized if (at least) one of the two matrices is positive definite (or negative definite). The conditions for this theorem are satisfied by the matrices V and T, which are both symmetric, and T is further positive definite. The proof of the theorem consists of constructing three matrices that describe three coordinate transformations: a rotation, a scale transformation, and another rotation.

To simplify the manipulations we use the matrix formalism. Writing

$$\lambda = \omega^2 \qquad (10.31)$$

for the eigenvalue, and defining the column matrix X by

$$X = \begin{pmatrix} a_1 \\ \vdots \\ a_f \end{pmatrix}, \qquad (10.32)$$

we can cast (10.16) into the form

$$(V - \lambda T)X = 0 \qquad (10.33)$$

Step 1. Let us first consider the eigenvalue equation

$$(T - \mu 1)Y = 0 \qquad (10.34)$$

for the matrix T alone. Because T is symmetric, its eigenvalues μ_i are real, and the associated eigenvectors $Y^{(i)}$ are either always orthogonal, if the eigenvalues are all different, or can be chosen orthogonal, if the eigenvalue spectrum is degenerate (see Section 2.12). The eigenvectors may also be chosen normalized such that

$$Y^{(i)T}Y^{(j)} = \delta_{ij} \tag{10.35}$$

The matrix

$$M = \left(Y^{(1)}, \ldots Y^{(f)}\right) \tag{10.36}$$

whose columns consist of the normalized eigenvectors $Y^{(i)}$ of T is orthogonal, because by (10.35),

$$M^T M = \begin{pmatrix} Y^{(1)T} \\ \vdots \\ Y^{(f)T} \end{pmatrix} \left(Y^{(1)} \ldots Y^{(f)}\right) = 1 = MM^T \tag{10.37}$$

M transforms T into the diagonal matrix

$$D = M^T T M \tag{10.38}$$

whose elements consist of the eigenvalues of T. This can be shown as follows. Multiplying the equation

$$(T - \mu_j 1)Y^{(j)} = 0$$

from the left by M^T, and applying (10.37) we obtain

$$M^T(T - \mu_j 1)1Y^{(j)} = M^T(T - \mu_j 1)MM^T Y^{(j)}$$

$$= \left(M^T T M - \mu_j M^T M\right)M^T Y^{(j)} = 0$$

or

$$(D - \mu_j 1)Z^{(j)} = 0 \tag{10.39}$$

where $Z^{(j)}$ is defined by

$$Z^{(j)} = M^T Y^{(j)} \tag{10.40}$$

By (10.35), the components of $Z^{(j)}$ are

$$\left(Z^{(j)}\right)_i = \left(M^T Y^{(j)}\right)_i = \delta_{ij},$$

such that (10.39) implies

$$D_{ij} = \mu_j \delta_{ij} \tag{10.41}$$

The transformation from T to $M^T TM$ is called a principal axis transformation. Because T_{ij} defines a positive quadratic form, the equation

$$\sum_{i=1}^{f} \sum_{j=1}^{f} T_{ij} a_i a_j = \text{constant} \ (> 0)$$

defines an f-dimensional ellipsoid in some arbitrary orientation to the coordinate axes. The vectors $Z^{(j)}$ point in the direction of the semiaxes of the ellipsoid, and the matrix M defines, therefore, a rotation from the given coordinate axes to the semiaxes (or principal axes) of the ellipsoid. We encountered this transformation previously in the discussion of the inertia tensor I (section 9.4.8). There, however, we did not make use of the full mathematical apparatus available because it was simple to visualize the geometrical properties of the principal axis transformation in three dimensions.

Step 2. Because T defines a positive definite quadratic form, we have

$$0 < Y^T TY = (Y^T M)(M^T TM)(M^T Y)$$

$$= (M^T Y)^T (M^T TM)(M^T Y) = Z^T DZ = \sum_{j=1}^{f} \mu_j Z_j^2$$

for any choice of real numbers Z_j (not all Z_j being equal to zero). The inequality implies that all eigenvalues μ_j of T are positive, and that $\det D > 0$. We can, therefore, define a diagonal matrix N by

$$N^2 = D^{-1} \tag{10.42}$$

There are, in general, 2^f different matrices N that satisfy this equation, and we choose for simplicity that N whose elements are all positive:

$$N = \begin{pmatrix} +\mu_1^{-1/2} & & 0 \\ & \ddots & \\ 0 & & +\mu_f^{-1/2} \end{pmatrix}$$

The matrix N defines a scale transformation of the coordinate axes. If N is applied to some vector Z, then the ith component of NZ is changed by the factor $\mu_i^{-1/2}$.

Applying the combined transformation NM to Eq. 10.31 we obtain

$$(NM^T VMN - \lambda NM^T TMN)(N^{-1} M^T X) = 0 \tag{10.43}$$

By definition of the matrices M and N, we find that

$$NM^T TMN = 1, \tag{10.44}$$

and that

$$U = NM^T VMN \tag{10.45}$$

is a symmetric matrix. Setting

$$W = N^{-1}M^T X, \tag{10.46}$$

and using (10.44) and (10.45), we arrive at

$$(U - \lambda 1)W = 0, \tag{10.47}$$

which is the desired eigenvalue equation in standard form. Note that the eigenvalue λ remained unchanged under the performed transformations.

Step 3. The last step consists of performing another principal axis transformation. Because U is symmetric (see Problem 10.9), the eigenvalues λ_j of U must be real, and the associated eigenvectors $W^{(j)}$ are orthogonal (or may be chosen orthogonal in the case of degeneracy). If we require that the $W^{(j)}$ are normalized, we obtain

$$\delta_{ij} = W^{(i)T}W^{(j)} = (N^{-1}M^T X^{(i)})^T (N^{-1}M^T X^{(j)})$$

$$= X^{(i)T}MN^{-1}N^{-1}M^T X^{(j)},$$

or because of (10.38) and (10.42),

$$X^{(i)T}TX^{(j)} = \delta_{ij} \tag{10.48}$$

These orthogonality conditions for the vectors $X^{(i)}$ read in coordinate form as

$$\sum_{k=1}^{f} \sum_{l=1}^{f} T_{kl}a_{ki}a_{lj} = \delta_{ij} \tag{10.49}$$

If we introduce the matrix

$$A = \left(X^{(1)} \ldots X^{(f)}\right) = \begin{pmatrix} a_{11} & \cdots & a_{1f} \\ \vdots & & \vdots \\ a_{f1} & \cdots & a_{ff} \end{pmatrix} \tag{10.50}$$

we can write the orthogonality condition also in the alternate form

$$A^T TA = 1 \tag{10.51}$$

The f^2 equations (10.48) or (10.49) are not all independent. If $i \neq j$, there are only $f(f-1)/2$ independent equations expressing orthogonality, which is a consequence of the symmetry property of the matrix T (or U). If $i = j$, there are f independent inhomogeneous equations representing normalization conditions for the coefficients a_{ki} or vectors $X^{(i)}$. We may call those solutions a_{ki} that satisfy Eq. 10.49, the normalized solutions. They differ from the solutions obtained in the preceding section by constant (scale) factors for each eigenvalue λ_i.

The constants D_i and E_i appearing in Eq. 10.21 can now be found in explicit form from the initial conditions (10.24) and (10.25), respectively. Multiplying Eq. 10.24 by $T_{kl}a_{ki}$ and summing over k and l, we obtain

$$\sum_{k=1}^{f} \sum_{l=1}^{f} T_{kl}a_{ki}\xi_{l_0} = \sum_{k=1}^{f} \sum_{l=1}^{f} T_{kl}a_{ki} \sum_{j=1}^{f} a_{lj}D_j$$

$$= \sum_{j=1}^{f} \left(\sum_{k=1}^{f} \sum_{l=1}^{f} T_{kl}a_{ki}a_{lj} \right) D_j$$

The term in parentheses is by Eq. 10.49 equal to δ_{ij} such that

$$D_i = \sum_{j=1}^{f} \delta_{ij}D_j = \sum_{k=1}^{f} \sum_{l=1}^{f} a_{ki}T_{kl}\xi_{l_0} \tag{10.52}$$

Similarly, for E_i we have

$$E_i = \frac{1}{\omega_i} \sum_{k=1}^{f} \sum_{l=1}^{f} a_{ki}T_{kl}v_{l_0} \tag{10.53}$$

It must not be forgotten that D_i and E_i can be written in this simple closed form only if the a_{ki} are the normalized solutions of (10.16) or (10.33).

Let us now introduce the orthogonal matrix

$$P = \left(W^{(1)} \ldots W^{(f)} \right) \tag{10.54}$$

whose columns consist of the orthonormal eigenvectors $W^{(j)}$ of U, see Eq. 10.47. Remembering Eqs. 10.46 and 10.50, we have also

$$P = N^{-1}M^{T}A \tag{10.55}$$

which will be needed later. P transforms U into the diagonal matrix $P^{T}UP$ whose elements are the eigenvalues of U:

$$(P^{T}UP)_{ij} = \lambda_i\delta_{ij} \tag{10.56}$$

(see the derivation of Eq. 10.41).

This completes the proof of the theorem that V and T can be simultaneously diagonalized. The remaining problem, namely, how the set of normal coordinates for the system of coupled oscillators can be obtained, will be discussed in the following section.

Table 10.1 Matrix Equations for Coupled Oscillators

Equation	$T_{ij}=T_{ji}$ (arbitrary)	$T_{ij} = m_i\,\delta_{ij}$	$T_{ij} = m\delta_{ij}$
10.33	$(V - \lambda T)X = 0$		
10.50	$A = (X^{(1)}\ldots X^{(f)})$		
10.51	$A^T TA = 1$	$A^T TA = 1$	$A^T A = m^{-1}1$
10.34	$(T - \mu 1)Y = 0$		
10.36	$M = (Y^{(1)}\ldots Y^{(f)})$	$M = 1$	$M = 1$
10.38, 10.42	$N^2 = (M^T TM)^{-1}$	$N_{ij} = m_i^{-1/2}\delta_{ij}$	$N = m^{-1/2}1$
10.44	$U = NM^T VMN$	$U = NVN$	$U = m^{-1}V$
10.46	$W = N^{-1}M^T X$	$W = N^{-1}X$	$W = \sqrt{m}\,X$
10.47	$(U - \lambda 1)W = 0$	$(U - \lambda 1)W = 0$	$(m^{-1}V - \lambda 1)W = 0$
10.54	$P = (W^{(1)}\ldots W^{(f)})$	$P = (\sqrt{m_1}\,X^{(1)}\ldots \sqrt{m_f}X^{(f)})$	$P = \sqrt{m}\,A$

The calculation of the matrix elements is often very tedious in practical applications. Fortunately, in two special cases the formalism can be greatly simplified. If Cartesian coordinates are used, the matrix T is diagonal. If, further, all masses are equal, $m_i = m$, the matrix T is a multiple of the unit matrix. The resulting equations are summarized in Table 10.1.

We conclude this section with two examples.

Example 10.7

Find the normalized solutions a_{ki} for the double pendulum described in Example 10.5.

The normalized solutions must be multiples of the special solutions (10.29) and (10.30), that is,

$$a_{11} = \alpha, \qquad a_{21} = \sqrt{2}\,\alpha$$

$$a_{12} = \beta, \qquad a_{22} = -\sqrt{2}\,\beta$$

Our problem consists of determining the factors α and β. The coefficients of T are found from (10.11):

$$T_{11} = 2ml^2, \qquad T_{12} = T_{21} = T_{22} = ml^2 \tag{10.57}$$

Using Eq. 10.49, we have for $i = 1$

$$\sum_{k=1}^{2}\sum_{l=1}^{2} T_{kl}a_{k1}a_{l1} = 1$$

or

$$T_{11}a_{11}a_{11} + T_{12}a_{11}a_{21} + T_{21}a_{21}a_{11} + T_{22}a_{21}a_{21}$$

$$= ml^2\alpha\left(2 + \sqrt{2} + \sqrt{2} + \sqrt{2}\sqrt{2}\right) = 1$$

such that

$$\alpha = \frac{1}{\sqrt{2}\sqrt{2 + \sqrt{2}}} \frac{1}{\sqrt{ml^2}}$$

Similarly, for $i = 2$, we obtain

$$\beta = \frac{1}{\sqrt{2}\sqrt{2 - \sqrt{2}}} \frac{1}{\sqrt{ml^2}}$$

The normalized solution is

$$A = \frac{1}{\sqrt{ml^2}} \begin{pmatrix} \left(2(2 + \sqrt{2})\right)^{-1/2} & \left(2(2 - \sqrt{2})\right)^{-1/2} \\ (2 + \sqrt{2})^{-1/2} & (2 - \sqrt{2})^{-1/2} \end{pmatrix} \tag{10.58}$$

The vectors

$$X^{(1)} = \begin{pmatrix} a_{11} \\ a_{21} \end{pmatrix} \quad \text{and} \quad X^{(2)} = \begin{pmatrix} a_{21} \\ a_{22} \end{pmatrix}$$

are orthogonal in the sense of Eq. 10.48 (see Problem 10.11). ∎

Example 10.8

Using the normalized solution (10.58), find the constants D_i and E_i for the double pendulum described in Example 10.5.

We apply the results (10.57) and (10.58) to Eqs. 10.52 and 10.53:

$$D_1 = \sum_{k=1}^{2}\sum_{l=1}^{2} a_{k1} T_{kl} \xi_{l_0}$$

$$= (a_{11}T_{11} + a_{21}T_{21})\xi_{1_0} + (a_{11}T_{12} + a_{21}T_{22})\xi_{2_0}$$

$$= \left((2 + \sqrt{2})ml^2\right)^{-1/2} ml^2 \left[\left(\frac{1}{\sqrt{2}}2 + 1\right)\xi_{1_0} + \left(\frac{1}{\sqrt{2}} + 1\right)\xi_{2_0}\right]$$

$$= \left(\frac{1}{2}(2 + \sqrt{2})ml^2\right)^{1/2}\left(\xi_{1_0} + \frac{1}{\sqrt{2}}\xi_{2_0}\right),$$

$$D_2 = \left(\frac{1}{2}(2 - \sqrt{2})ml^2\right)^{1/2}\left(\xi_{1_0} - \frac{1}{\sqrt{2}}\xi_{2_0}\right),$$

$$E_1 = \frac{1}{\omega_1}\left(\frac{1}{2}(2 + \sqrt{2})ml^2\right)^{1/2}\left(\nu_{1_0} + \frac{1}{\sqrt{2}}\nu_{2_0}\right),$$

$$E_2 = \frac{1}{\omega_2}\left(\frac{1}{2}(2 - \sqrt{2})ml^2\right)^{1/2}\left(\nu_{1_0} - \frac{1}{\sqrt{2}}\nu_{2_0}\right)$$

It is obvious from Eqs. 10.52 and 10.53 that the way the a_{ki} are normalized affects the constants D_i and E_i. The results obtained here are necessarily different from those obtained in Example 10.6. However, the solutions $\xi_i(t)$ remain unaffected by the normalization conditions. Evaluation of the right side of Eq. 10.21 with the normalized coefficients (10.58) leads in fact to the same results that were derived in Example 10.6 (see Problem 10.12).

Our example also shows that a formally simpler procedure need not automatically imply less work in actual calculations. ■

10.6 NORMAL COORDINATES

A normal coordinate $\zeta_i(t)$ is a generalized coordinate $\xi_i(t)$ that depends on the normal frequency ω_i only, or, in other words, whose time dependence is of the form (10.26). We wish to find a transformation that yields the vector[2]

$$Z = \begin{pmatrix} \zeta_1(t) \\ \vdots \\ \zeta_f(t) \end{pmatrix} \tag{10.59}$$

directly from the vector

$$\Xi = \begin{pmatrix} \xi_1(t) \\ \vdots \\ \xi_f(t) \end{pmatrix} \tag{10.60}$$

The matrix P, Eq. 10.54, transforms the eigenvectors $W^{(j)}$ belonging to $\lambda_j = \omega_j^2$ into orthonormal vectors $P^T W^{(j)}$ whose jth component is equal to one, whereas all other components are zero. The vectors $P^T W^{(j)}$ can therefore be chosen as basis vectors whereby each basis vector is associated with one frequency ω_j, and the normal coordinates are then the coordinates of any vector Ξ with respect to this new basis. Thus, with the matrices M^T (rotation), N^{-1} (scale transformation) and P^T (another rotation), we have

$$Z = P^T N^{-1} M^T \Xi \tag{10.61}$$

This equation is, however, not very convenient in practical calculations because one would first have to calculate the three transformation matrices, which is not an easy task in general. Fortunately, Eq. 10.61 can be reduced to a much simpler expression by applying the algebraic equations (10.55), (10.38), and (10.42):

$$Z = (N^{-1} M^T A)^T N^{-1} M^T \Xi$$

$$= A^T M N^{-1} N^{-1} M \Xi$$

[2]Z defined by Eq. 10.59 is different from that Z introduced briefly by Eq. 10.40 for mathematical convenience.

or

$$Z = A^T T \Xi \tag{10.62}$$

In coordinates this reads

$$\zeta_i(t) = \sum_{j=1}^{f} \sum_{k=1}^{f} a_{ji} T_{jk} \xi_k(t)$$

$$= \sum_{j=1}^{f} \sum_{k=1}^{f} \xi_k(t) T_{kj} a_{ji} \tag{10.63}$$

because $T_{jk} = T_{kj}$. Applying Eq. 10.51 to Eq. 10.62, we easily find Ξ from Z:

$$Z = A^T T 1 \Xi = A^T T A A^{-1} \Xi = A^{-1} \Xi$$

or

$$\Xi = AZ \tag{10.64}$$

This equation is often used to define the normal coordinates.

Both the potential energy and the kinetic energy assume simple forms in terms of the normal coordinates. The potential energy can be written in matrix form

$$V = \tfrac{1}{2} \Xi^T V \Xi \tag{10.65}$$

In the system of normal coordinates, the matrix V is diagonal, and Eqs. 10.56 and 10.65 become

$$V = \frac{1}{2} \sum_{i=1}^{f} \sum_{j=1}^{f} \zeta_i \lambda_i \delta_{ij} \zeta_j$$

$$= \frac{1}{2} \sum_{i=1}^{f} \lambda_i \zeta_i^2 \tag{10.66}$$

The kinetic energy reads in matrix form

$$T = \tfrac{1}{2} \dot{\Xi}^T T \dot{\Xi} \tag{10.67}$$

In the system of normal coordinates, the matrix T is the unit matrix $1 = P^T 1 P$ by Eqs. 10.44 and 10.54, and (10.67) reads

$$T = \frac{1}{2} \sum_{i=1}^{f} \sum_{j=1}^{f} \dot{\zeta}_i \delta_{ij} \dot{\zeta}_j$$

$$= \frac{1}{2} \sum_{i=1}^{f} \dot{\zeta}_i^2 \tag{10.68}$$

V and T reduce to sums of squares of the normal coordinates and their time derivatives, respectively.

If the Lagrangian is written in terms of the normal coordinates and velocities, we have

$$L = T - V = \frac{1}{2} \sum_{i=1}^{f} \left(\dot{\zeta}_i^2 - \lambda_i \zeta_i^2 \right) \tag{10.69}$$

Lagrange's equations are then

$$\Lambda_j L = \ddot{\zeta}_j + \lambda_j \zeta_j = 0, \qquad j = 1, \ldots f \tag{10.70}$$

which have

$$\zeta_j(t) = A_{j_+} e^{i\omega_j t} + A_{j_-} e^{-i\omega_j t}$$

$$= A_j \cos \omega_j t + B_j \sin \omega_j t \tag{10.71}$$

as solutions. Whereas (at least in most practical applications) the coordinate ξ_l relates to one mass point that undergoes oscillation with all normal frequencies ω_j simultaneously, the normal coordinate ζ_j describes oscillation of the whole system with only one frequency ω_j.

Example 10.9

Find the normal coordinates of the double pendulum described in Example 10.5.
Combining Eqs. 10.57, 10.58, and 10.62, we obtain

$$\begin{pmatrix} \zeta_1 \\ \zeta_2 \end{pmatrix} = \frac{1}{\sqrt{ml^2}} \begin{pmatrix} \left(2(2 + \sqrt{2})\right)^{-1/2} & (2 + \sqrt{2})^{-1/2} \\ \left(2(2 - \sqrt{2})\right)^{-1/2} & (2 - \sqrt{2})^{-1/2} \end{pmatrix} ml^2 \begin{pmatrix} 2 & 1 \\ 1 & 1 \end{pmatrix} \begin{pmatrix} \xi_1 \\ \xi_2 \end{pmatrix}$$

The normal coordinates are

$$\zeta_1 = \sqrt{ml^2 \left(1 + 1/\sqrt{2}\right)} \left(\xi_1 + \xi_2/\sqrt{2} \right)$$

$$\zeta_2 = \sqrt{ml^2 \left(1 - 1/\sqrt{2}\right)} \left(\xi_1 - \xi_2/\sqrt{2} \right)$$

It is left as an exercise to show that ζ_1 and ζ_2 indeed satisfy Eqs. 10.67 and 10.68. ∎

10.7 THE WEIGHTED STRING

A weighted string is a system of particles that is attached to a massless elastic string having its ends fastened to rigid walls. The investigation of the motion of the weighted string is instructive for several reasons. First, the weighted string serves as a one-dimensional model for elastic bodies. The knowledge of the modes of oscillation of the particles is important for understanding some of the macroscopic properties of the solid, specific heats for example. Second, the equations of motion are exactly solvable with little mathematical effort for any number of particles if the system exhibits simplifying symmetry properties. Third, the methods of finding the modes of oscillation of the weighted string can easily be extended to similar two- and three-dimensional systems and to systems with continuous mass distributions. Fourth, the concept of boundary conditions (the ends of the string are held fixed) is introduced for the first time. Boundary conditions play the same role in the theory of partial differential equations (needed to describe continuous mass distributions or fields) as do initial conditions in the theory of ordinary differential equations (needed to describe particles): They single out the actual solution from the infinitely many possible solutions.

To be more specific, let us consider N particles, each of mass m, attached at equal intervals of size d to an elastic massless string of length $(N + 1)d$ held under the tension F. The string is aligned along the x-axis of a Cartesian coordinate system. If we allow the particles to oscillate in the x-y-plane, the general motion of the particles consists of a superposition of longitudinal and transverse oscillations about their equilibrium points. The coordinates of the particles in equilibrium are

$$x_l = ld, \qquad y_l = 0 \quad \text{for} \quad l = 1, 2, \ldots N \qquad (10.72)$$

It is convenient to denote the coordinates of the endpoints of the string by

$$x_0 = 0, \qquad y_0 = 0 \quad \text{and} \quad x_{N+1} = (N + 1)d, \qquad y_{N+1} = 0 \quad (10.73)$$

In the case of longitudinal motion (along the x-axis) each segment of the string may be considered as a small spring with spring constant k. The potential energy of the $(l + 1)$st segment between the lth and $(l + 1)$st mass point is

$$\tfrac{1}{2}k(x_{l+1} - x_l)^2, \qquad (10.74)$$

(see Figure 10.7), and the potential energy V_{long} of the string for longitudinal motion is the sum of all $(N + 1)$ contributions (10.74),

$$V_{\text{long}}(x) = \sum_{l=0}^{N} \frac{1}{2}k(x_{l+1} - x_l)^2 \qquad (10.75)$$

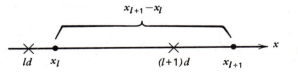

Figure 10.7 Weighted string — longitudinal displacement.

In the case of transverse motion (in the y-direction), the relative distance between the lth and the $(l + 1)$st particle is

$$D_l = \sqrt{(y_{l+1} - y_l)^2 + d^2} = d\sqrt{1 + \frac{(y_{l+1} - y_l)^2}{d^2}}$$

$$= d\left[1 + \frac{1}{2}\left(\frac{y_{l+1} - y_l}{d}\right)^2 + \cdots\right]$$

(see Figure 10.8). If only terms up to second order in the coordinates are kept, the amount by which the lth segment is stretched is given by

$$D_l - d = \frac{1}{2d}(y_{l+1} - y_l)^2 \tag{10.76}$$

The potential energy is by Eq. 10.5 a homogeneous quadratic form in the coordinates. This implies together with (10.76) that the increase ΔF of the tension, caused by the transverse displacement of the string, does not contribute to the potential energy in this approximation. We then have $F = $ constant, and the potential energy V_{trans} for transverse displacement becomes

$$V_{\text{trans}}(y_l) = \sum_{l=0}^{N} F(D_l - d) = \frac{F}{2d}\sum_{l=0}^{N}(y_{l+1} - y_l)^2 \tag{10.77}$$

We see that V_{long} and V_{trans} have the same mathematical structure.

The kinetic energies for longitudinal and transverse motion of the system are given by

$$T_{\text{long}}(\dot{x}_l) = \sum_{l=1}^{N} \frac{1}{2} m\dot{x}_l^2 \tag{10.78}$$

and

$$T_{\text{trans}}(\dot{y}_l) = \sum_{l=1}^{N} \frac{1}{2} m\dot{y}_l^2, \tag{10.79}$$

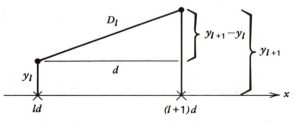

Figure 10.8 Weighted string — transverse displacement.

respectively. Because the Lagrangians

$$L_{\text{long}}(x_l, \dot{x}_l) = T_{\text{long}} - V_{\text{long}} \tag{10.80}$$

and

$$L_{\text{trans}}(y_l, \dot{y}_l) = T_{\text{trans}} - V_{\text{trans}} \tag{10.81}$$

depend on different coordinates, it follows from Lagrange's equations that the longitudinal and transverse modes are independent of each other. Further, because the Lagrangians (10.80) and (10.81) have the same mathematical form, the equations of motion and hence their general solutions must be the same. It is sufficient, therefore, to investigate only one form of the motion. We choose the transverse mode because it can more easily be visualized.

From (10.77) we obtain the force F_l on the lth particle:

$$F_l = -\frac{\partial V}{\partial y_l} = -\frac{\partial}{\partial y_l} \frac{F}{2d} \left(\cdots + (y_l - y_{l-1})^2 + (y_{l+1} - y_l)^2 + \cdots \right)$$

$$= -\frac{F}{d}(-y_{l-1} + 2y_l - y_{l+1}) \tag{10.82}$$

Performing another differentiation we see that the only nonvanishing matrix elements V_{kl} of V are

$$V_{kk} = \frac{2F}{d}, \qquad V_{kk+1} = V_{kk-1} = -\frac{F}{d}$$

such that V has the form

$$V = \frac{F}{d} \begin{pmatrix} 2 & -1 & & & & \\ -1 & 2 & -1 & & 0 & \\ & -1 & 2 & -1 & & \\ & & -1 & & \ddots & \\ & 0 & & & & \\ & & & & -1 & 2 \end{pmatrix} \tag{10.83}$$

The matrix T is a multiple of the unit matrix (see Table 10.1):

$$T = m\mathbf{1} \tag{10.84}$$

It turns out that the formalism developed in the preceding sections is at this level of little practical use for our example because solving the characteristic equation

$$\det(V - m\lambda\mathbf{1}) = 0$$

with V given by (10.83) for arbitrary N is rather difficult. Fortunately, the normal frequencies can be found directly from the equations of motion

$$m\ddot{y}_l = F_l = -\frac{F}{d}(-y_{l-1} + 2y_l - y_{l+1}), \qquad l = 1 \ldots N \qquad (10.85)$$

Here, Newton's and Lagrange's equations are identical because we have chosen Cartesian coordinates as the generalized coordinates. The system (10.84) of N coupled differential equations can be transformed into a system of N algebraic equations if we substitute trial functions of the form (10.14),

$$y_l = Ca_l\cos(\omega t + \delta), \qquad l = 1 \ldots N \qquad (10.86)$$

into Eq. 10.85. The result is, after rearrangement of terms and division by common factors,

$$\left(\frac{md}{2F}\omega^2 - 1\right)2a_l + a_{l-1} + a_{l+1} = 0 \qquad (10.87)$$

This equation may be solved if we set

$$a_l = a \sin lp\pi \qquad (10.88)$$

where p is some number to be determined. Combining (10.87) and (10.88), we obtain

$$\left(\frac{md}{2F}\omega^2 - 1\right)2a \sin lp\pi + a\sin(l-1)p\pi + a\sin(l+1)p\pi = 0$$

Applying addition theorems for the sine functions and solving for ω^2, we find

$$\omega^2 = \frac{2F}{md}(1 - \cos lp\pi) \qquad (10.89)$$

It seems now that ω^2 may assume any value between 0 and $4F/md$, but this is not the case. We have not yet applied the boundary conditions (10.73) to y_0 and y_{N+1}, that is, to a_0 and a_{N+1}. For $l = 0$, the boundary condition is identically satisfied:

$$0 = a_0 = a \sin 0p\pi = 0$$

For $l = N + 1$, we obtain

$$0 = a_{N+1} = a\sin(N+1)p\pi$$

and the values of p are restricted to

$$(N+1)p\pi = j\pi,$$

where j is an integer. Thus,

$$p = \frac{j}{N+1} \qquad (10.90)$$

and (10.88) becomes

$$\omega^2 = \frac{2F}{md}\left(1 - \cos\frac{j\pi}{N+1}\right) = \omega_j^2 \qquad (10.91)$$

We wish to determine the number of normal frequencies ω_j, or, equivalently, the range of integers j. Because the arc cosine function is uniquely defined on the interval $[-1, 1]$, j may assume only the $N + 2$ values $0, 1, \ldots N, N + 1$. Also, because the cosine function is monotonically decreasing on the interval $[0, \pi]$, all frequencies ω_j are different. However, if $j = 0$, we have $\omega_0^2 = 0$, and the system must be in equilibrium. If $j = N + 1$, then $p = 1$; and, although we would have $\omega_{N+1}^2 = 4F/md \neq 0$, Eq. 10.88 yields $a_l = 0$ for all l. Again, the system must be in equilibrium. Hence, there are exactly N different normal frequencies ω_j of the form (10.91), with j ranging from 1 through N. We know already from the preceding sections that there are at most N (different) normal frequencies, but this result followed here directly from the boundary conditions. For $N = 4$, the frequency spectrum is shown in Figure 10.9.

For large numbers of particles, the lowest order frequencies may be approximated by

$$\omega_j^2 = \frac{2F}{md}\left[1 - \left(1 - \frac{1}{2!}\left(\frac{j\pi}{N+1}\right)^2 + \ldots\right)\right]$$

such that

$$\omega_j = \sqrt{\frac{F}{md}}\,\frac{\pi}{N+1}j$$

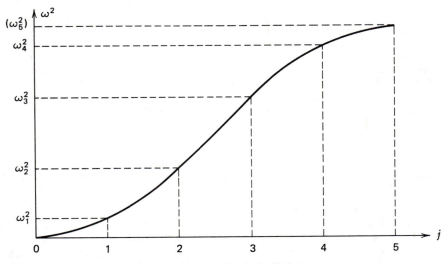

Figure 10.9 Frequency spectrum of a string loaded with four masses.

These frequencies are integral multiples of the fundamental frequency

$$\omega_1 = \sqrt{\frac{F}{md}} \frac{\pi}{N+1}$$

and are called the jth harmonics of ω_1.

The unnormalized normal modes are found by substituting p from Eq. 10.90 into (10.88):

$$a_{lj} = a_j \sin \frac{lj\pi}{N+1} \tag{10.92}$$

We have attached the index j to both a_l and a to express their dependence on the normal frequency ω_j. In order to normalize the coefficients a_{lj} we substitute (10.84) and (10.92) into (10.49), and, setting $i = j$, we arrive at the condition

$$1 = m \sum_{k=1}^{N} a_{kj}a_{lj} = ma_j^2 \sum_{k=1}^{N} \sin^2 \frac{kj\pi}{N+1} \tag{10.93}$$

Extending the sum to $N + 1$, because $\sin j\pi = 0$, and using the trigonometric identity $\sin^2 x = (1 - \cos 2x)/2$, we may write (10.93) in the form

$$1 = ma_j^2 \sum_{k=1}^{N+1} \frac{1}{2} \left(1 - \cos \frac{2kj\pi}{N+1} \right)$$

$$= ma_j^2 \left[\frac{1}{2}(N+1) - \frac{1}{2} \sum_{k=1}^{N+1} \cos \frac{2kj\pi}{N+1} \right] \tag{10.94}$$

The sum in (10.94) vanishes identically (see Problem 10.14), and the constants

$$a_j = \left(\frac{2}{m(N+1)} \right)^{1/2} \tag{10.95}$$

turn out to be independent of j. The normalized coefficients a_{lj} are then given by

$$a_{lj} = \left(\frac{2}{m(N+1)} \right)^{1/2} \sin \frac{lj\pi}{N+1} \tag{10.96}$$

The particle index l and the frequency index j, both running from 1 through N, enter Eq. 10.96 in a symmetrical way, and the matrix P, Eq. 10.54, is not only orthogonal but also symmetric.

In order to get a better understanding of the normal modes, let us consider the case $N = 4$. Figure 10.10 shows the graphs of the five functions $\sin(jl\pi/5)$ for $j = 1, \ldots 5$ over the interval $0 \leqslant l \leqslant 5$. The values of these functions at $l = 1, 2, 3,$ and 4 are the normal modes a_{lj}. At a given instant, let "$+$" stand for the motion of a particle in the positive y-direction, "$-$" for the motion in the

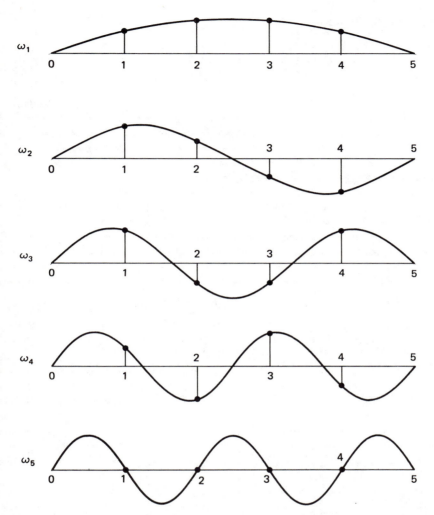

Figure 10.10 Normal modes of a string loaded with four masses.

negative y-direction. The following modes of oscillation are possible in this case:

$$\omega_1 = 0.62\sqrt{F/md} : \quad + \ + \ + \ +$$
$$\omega_2 = 1.18\sqrt{F/md} : \quad + \ + \ - \ -$$
$$\omega_3 = 1.62\sqrt{F/md} : \quad + \ - \ - \ +$$
$$\omega_4 = 1.90\sqrt{F/md} : \quad + \ - \ + \ -$$

For $\omega_5 = 2\sqrt{F/md}$, all normal modes are zero as expected.

The general solution $y_l(t)$, Eq. 10.86, reads

$$y_l(t) = \sum_{j=1}^{N} a_{lj} C_j \cos(\omega_j t + \delta_j) = \sqrt{\frac{2}{m(N+1)}} \sum_{j=1}^{N} C_j \sin\frac{lj\pi}{N+1} \cos(\omega_j t + \delta_j)$$

$$(10.97)$$

where the constants C_j and δ_j can be calculated from the initial values of y_l and $\dot{y}_l$.

10.8 THE CONTINUOUS STRING

The continuous (massive) string may be considered as a weighted string consisting of infinitely many particles whose total mass remains finite. The equation of motion for the continuous string can be derived from the equations of motion (10.85) for the loaded string if the following substitutions are made:

particle index l	$\rightarrow$	string coordinate x
particle displacement $y_l(t)$	$\rightarrow$	continuous function $y(x, t)$
interval size d	$\rightarrow$	differential dx
mass m of a particle	$\rightarrow$	differential $dm = \lambda\, dx$, where
		λ is the linear mass density
ordinary derivatives	$\rightarrow$	partial derivatives

Equation 10.85 stated in the form

$$m\frac{d^2 y_l(t)}{dt^2} = F\left[\frac{y_{l+1}(t) - y_l(t)}{d} - \frac{y_l(t) - y_{l-1}(t)}{d}\right]$$

then becomes

$$\lambda\, dx \frac{\partial^2 y(x, t)}{\partial t^2} = F\left[\frac{\partial y(x + dx, t)}{\partial x} - \frac{\partial y(x, t)}{\partial x}\right]$$

$$= F\left[\frac{\partial y(x, t)}{\partial x} + \frac{\partial^2 y(x, t)}{\partial x^2}\, dx - \frac{\partial y(x, t)}{\partial x}\right]$$

$$= F\frac{\partial^2 y(x, t)}{\partial x^2}\, dx$$

from which follows

$$\frac{\partial^2 y}{\partial x^2} - \frac{1}{\left(\frac{F}{\lambda}\right)}\frac{\partial^2 y}{\partial t^2} = 0 \qquad (10.98)$$

Infinitely many ordinary differential equations for particle motion have been replaced by one partial differential equation for continuous mass distributions.

The equation

$$\nabla^2 f - \frac{1}{v^2} \frac{\partial^2 f}{\partial t^2} = 0$$

is called the three-dimensional wave equation, and the equation of motion (10.98) for the continuous string is therefore the one-dimensional wave equation.

We do not intend to develop the general mathematical formalism for solving the wave equation. But the solution to (10.98) is easily found if $F/\lambda = $ constant and if the string is held fixed at its ends, that is, the string is subjected to the boundary conditions

$$y(0, t) = 0 \qquad (10.99)$$

and

$$y(L, t) = 0 \qquad (10.100)$$

Assuming that the oscillations are of the form

$$y(x, t) = f(x)\cos(\omega t + \delta) \qquad (10.101)$$

we obtain from (10.98) the equation

$$\frac{d^2 f}{dx^2} + \frac{\omega^2}{\left(\frac{F}{\lambda}\right)} f = 0$$

or

$$f'' + k^2 f = 0 \qquad (10.102)$$

where we have set

$$k^2 = \frac{\lambda}{F}\omega^2 \qquad (10.103)$$

Equation 10.102 is called the Helmholtz equation associated with the wave equation (10.98). We recognize its similarity to the equation of motion of the simple harmonic oscillator. The general solution of (10.102) is

$$f(x) = A \sin kx + B \cos kx \qquad (10.104)$$

Applying the boundary condition (10.99) to (10.104), we obtain

$$f(0) = A \sin k0 + B \cos k0 = 0$$

which implies $B = 0$. And subjecting (10.104) to (10.100), we find

$$f(L) = A \sin kL = 0,$$

which is possible only if

$$k = \frac{j\pi}{L}, \qquad j = 1, 2, 3, \ldots$$

It follows that ω is limited to

$$\omega_j = \sqrt{\frac{F}{\lambda}} \frac{\pi}{L} j, \qquad j = 1, 2, 3, \ldots \tag{10.105}$$

All frequencies are harmonics of the fundamental frequency

$$\omega_1 = \sqrt{\frac{F}{\lambda}} \frac{\pi}{L}$$

This fact is of great significance to string instruments because the traditional pleasantly sounding harmonies in music consist of spectra of harmonic frequencies.

The general solution is given by

$$y(x, t) = \sum_{j=1}^{\infty} A_j \sin \frac{\pi}{L} jx \cos(\omega_j t + \delta_j), \tag{10.106}$$

which is a superposition of infinitely many standing waves. The similarity between the solutions (10.97) for the loaded string and (10.106) for the continuous string is obvious.

The longest day will have an end.

ANONYMOUS

PROBLEMS

10.1 (a) Find in terms of T_{kl} and V_{kl} the general expression for the normal frequencies of a two particle system.

(b) Simplify your answer for the case $T_{12} = 0$.

10.2 Three particles of mass m each are coupled in such a way that the potential energy is given by

$$V = \tfrac{1}{2}\left[k_1\left(x_1^2 + x_3^2\right) + k_2 x_2^2 + k_3\left(x_1 x_2 + x_2 x_3\right)\right]$$

where $k_3 = (2k_1 k_2)^{1/2}$.

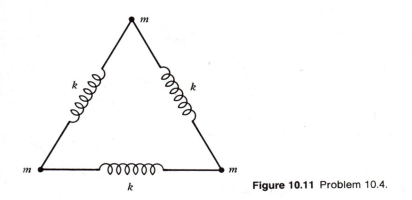

Figure 10.11 Problem 10.4.

(a) Find the point or points of equilibrium and determine their stability.

(b) Find the normal frequencies of the system.

10.3 Find the normal frequencies of a double pendulum for which $l_1 = l_2 = l$ if

(a) $m_1 \ll m_2$

(b) $m_1 \gg m_2$

10.4 Three particles connected by elastic springs are allowed to oscillate in the x-y plane (see Figure 10.11).

(a) Show that the normal frequencies are

$$\omega_1 = \omega_2 = 0, \qquad \omega_3 = 0, \qquad \omega_4 = \omega_5 = \sqrt{3k/2m}, \qquad \omega_6 = \sqrt{3k/m}$$

(b) *Without* solving any equations, guess the modes of oscillation (i.e., the directions of the motion of the particles at a given instant) associated with these frequencies. Do you expect a third mode associated with $\omega = \sqrt{3k/2m}$? Why is this mode redundant?

10.5 Find the normal frequencies and normal coordinates for the system shown in Figure 10.12.

10.6 Find the normal frequencies and normal modes for the linear triatomic molecule shown in Figure 10.5.

10.7 Find the normal frequencies and normal modes for the system shown in Figure 10.13.

10.8 Find the normal frequencies and normal modes of the coupled pendulum (Figure 10.3).

10.9 Show that the matrix U (Eq. 10.45) is symmetric.

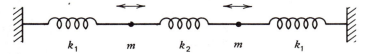

$$k_1 \qquad m \qquad k_2 \qquad m \qquad k_1$$

Figure 10.12 Problem 10.5.

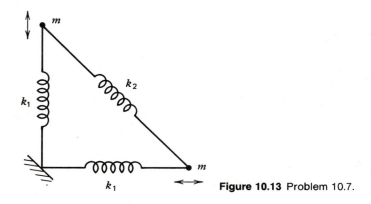

Figure 10.13 Problem 10.7.

10.10 Verify all equations in Table 10.1 for $T_{ij} = m_i \delta_{ij}$ and $T_{ij} = m \, \delta_{ij}$.

10.11 Show that the vectors $X^{(1)}$ and $X^{(2)}$ of Example 10.7 are orthogonal in the sense of Eq. 10.48.

10.12 Find the solution $\xi_i(t)$ for the double pendulum of Example 10.5 using the normalized coefficients (10.58).

10.13 Find V and T from Eqs. 10.66 and 10.68, respectively, using the results of Example 10.9. Compare your answer with the results obtained in Examples 10.1 and 10.3.

10.14 Show that

$$\sum_{k=1}^{N+1} \cos \frac{2kj\pi}{N+1} = 0$$

(see Eq. 10.94). Apply Euler's formula $\cos x = \frac{1}{2}(e^{ix} + e^{-ix})$ and the sum formula for the geometric series.

10.15 Find the normal coordinates for the weighted string.

10.16 (a) Derive Hamilton's equations for the weighted string using normal coordinates.

(b) Try to derive Hamilton's equations for the weighted string in arbitrary generalized coordinates. You will find out that Hamilton's formalism is not suitable to describe the motion of the weighted string.

10.17 Show that successive particles of a weighted string oscillate in opposite phases $(+ - + - + \ldots)$ in the highest frequency mode.

appendix

1 Vector Rules

Uniqueness: $\mathbf{a} = \sum_{i=1}^{N} a_i \mathbf{e}_i$ for a given set of basis vectors $\{\mathbf{e}_i\}$, $i = 1, 2, \ldots N$

Equality: $\mathbf{a} = \mathbf{b}$ if $a_i = b_i$ for all $i = 1 \cdots N$

Addition: $\mathbf{c} = \mathbf{a} \pm \mathbf{b}$ if $c_i = a_i \pm b_i$ for all $i = 1 \cdots N$

Multiplication by a scalar: $\mathbf{c} = \alpha \mathbf{a}$ if $c_i = \alpha a_i$ for all $i = 1 \cdots N$

Commutative law: $\mathbf{a} + \mathbf{b} = \mathbf{b} + \mathbf{a}$

Associative law: $\mathbf{a} + (\mathbf{b} + \mathbf{c}) = (\mathbf{a} + \mathbf{b}) + \mathbf{c}$

Distributive laws:

$$\alpha(\mathbf{a} \pm \mathbf{b}) = \alpha\mathbf{a} \pm \alpha\mathbf{b}$$

$$(\alpha \pm \beta)\mathbf{a} = \alpha\mathbf{a} \pm \beta\mathbf{a}$$

Inner (or scalar) product:

$$\mathbf{a} \cdot \mathbf{b} = ab \cos \sphericalangle(\mathbf{a}, \mathbf{b})$$

$$\mathbf{a} \cdot \mathbf{b} = \mathbf{b} \cdot \mathbf{a}$$

In an orthonormal coordinate system: $\mathbf{a} \cdot \mathbf{b} = \sum_{i=1}^{N} a_i b_i$

Magnitude of a vector: $a = \sqrt{\mathbf{a} \cdot \mathbf{a}}$

Outer (or vector) product (in three dimensions only):

$\mathbf{c} = \mathbf{a} \times \mathbf{b}$ where **(1)** $c = ab \sin \sphericalangle(\mathbf{a}, \mathbf{b})$

(2) $\mathbf{c}$ is perpendicular to both $\mathbf{a}$ and $\mathbf{b}$

(3) $\mathbf{a}$, the component of $\mathbf{b}$ perpendicular to $\mathbf{a}$, and $\mathbf{c}$ form a right-handed triad

Anticommutativity: $\mathbf{a} \times \mathbf{b} = -\mathbf{b} \times \mathbf{a}$

Representation in a Cartesian coordinate system:

$$\mathbf{c} = \begin{vmatrix} \mathbf{i} & \mathbf{j} & \mathbf{k} \\ a_x & a_y & a_z \\ b_x & b_y & b_z \end{vmatrix}$$

Multiple products (in three dimensions):

$$(\mathbf{abc}) = \begin{vmatrix} a_1 & a_2 & a_3 \\ b_1 & b_2 & b_3 \\ c_1 & c_2 & c_3 \end{vmatrix} = \mathbf{a} \cdot (\mathbf{b} \times \mathbf{c}) = \mathbf{b} \cdot (\mathbf{c} \times \mathbf{a}) = \mathbf{c} \cdot (\mathbf{a} \times \mathbf{b})$$

$$\mathbf{a} \times (\mathbf{b} \times \mathbf{c}) = \mathbf{b}(\mathbf{a} \cdot \mathbf{c}) - \mathbf{c}(\mathbf{a} \cdot \mathbf{b}) \quad \text{Mnemonic: } BAC - CAB \text{ rule}$$

$$(\mathbf{a} \times \mathbf{b}) \cdot (\mathbf{c} \times \mathbf{d}) = (\mathbf{a} \cdot \mathbf{c})(\mathbf{b} \cdot \mathbf{d}) - (\mathbf{a} \cdot \mathbf{d})(\mathbf{b} \cdot \mathbf{c})$$

$$(\mathbf{a} \times \mathbf{b}) \times (\mathbf{c} \times \mathbf{d}) = (\mathbf{abd})\mathbf{c} - (\mathbf{abc})\mathbf{d}$$

appendix

2 Special Classes of Forces and Associated Conserved Quantities

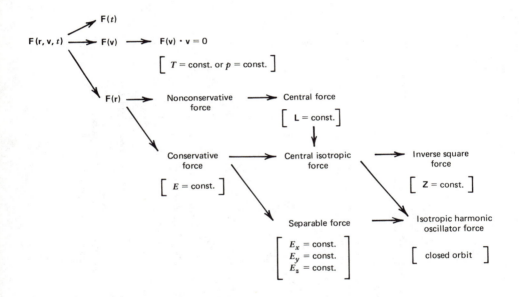

appendix

3

The Relationships Between the Various Formalisms of Classical Mechanics[a]

Formalism	Coordinates	Characteristic Function	Equations of Motion
Newton	$\mathbf{r}_i, \dot{\mathbf{r}}_i; i = 1 \cdots N$ N = number of particles	$V(\mathbf{r}_1, \cdots \mathbf{r}_N)$	$m_i \ddot{\mathbf{r}}_i = -\nabla_i V$
→	Projection from the $3N$-dimensional coordinate space $\langle \mathbf{r}_i \rangle$ to an f-dimensional coordinate space $\langle q_j \rangle$.		
Lagrange	$q_j, \dot{q}_j; j = 1 \cdots f$ f = degree of freedom	$L(q_j, \dot{q}_j) = T - V$	$\dfrac{d}{dt}\dfrac{\partial L}{\partial \dot{q}_j} - \dfrac{\partial L}{\partial q_j} = 0$
→	Legendre transformation from the $2f$-dimensional space $\langle q_j, \dot{q}_j \rangle$ to the $2f$-dimensional phase space $\langle q_j, p_j \rangle$.		
Hamilton	$q_j, p_j = \dfrac{\partial L}{\partial \dot{q}_j}; j = 1 \cdots f$	$H(q_j, p_j) = \displaystyle\sum_{j=1}^{f} \dot{q}_j \dfrac{\partial L}{\partial \dot{q}_j} - L$	$\dot{q}_j = \dfrac{\partial H}{\partial p_j}, \dot{p}_j = -\dfrac{\partial H}{\partial q_j}$
→	Canonical transformation in the $2f$-dimensional phase space[b]		
Hamilton	$Q_j, P_j; j = 1 \cdots f$	$K(Q_j, P_j) = H$	$\dot{Q}_j = \dfrac{\partial K}{\partial P_j}, \dot{P}_j = -\dfrac{\partial K}{\partial Q_j}$
→	Special canonical transformation in the $2f$-dimensional phase space		
Hamilton-Jacobi	$Q_j, P_j; j = 1 \cdots f$	$K(P_j) = H$	$\dot{Q}_j = \dfrac{\partial K}{\partial P_j} = \text{constant}, \dot{P}_j = \dfrac{\partial K}{\partial Q_j} = 0$

[a] This table applies to conservative systems subjected to time independent constraints only.

[b] A canonical transformation is a transformation of a set of coordinates $\langle q_j, p_j \rangle$ into another set of coordinates $\langle Q_k(q_j, p_j), P_k(q_j, p_j) \rangle$ such that Hamilton's equations retain their form in the new coordinates. A thorough discussion of the properties of canonical transformations and of the Hamilton-Jacobi theory goes beyond the scope of this book.

appendix
4 Axiomatic Formulations of Classical and Quantum Mechanics

The dynamical laws of classical mechanics can be introduced axiomatically without direct reference to Newton's laws. A possible set of axioms for classical mechanics is listed below and compared with a set of axioms for quantum mechanics.[1] It is tacitly assumed that an inertial frame exists. Observe the algebraic similarities of classical and quantum mechanics. Also note that no statements concerning the outcome of a measurement of an observable physical quantity have been made in Newtonian mechanics. With the advent of quantum mechanics, physicists were compelled to investigate the concept of measurement more thoroughly. The outcome of these investigations has greatly influenced the philosophy of knowledge (epistemology).

CLASSICAL MECHANICS

(1) There is a one-to-one mapping of any observable $\mathscr{F}$ on a real measurable function F in the phase space $\mathscr{P}$.

(2) The observables form a ring on which the Poisson bracket can be defined with the following properties:

(a) $\displaystyle (F, G) = \sum_{j=1}^{f} \left(\frac{\partial F}{\partial q_j} \frac{\partial G}{\partial p_j} - \frac{\partial G}{\partial q_j} \frac{\partial F}{\partial p_j} \right)$
$= -(G, F)$

(b) $(F, \alpha G) = \alpha(F, G)$

(c) $(F, G_1 + G_2) = (F, G_1) + (F, G_2)$

QUANTUM MECHANICS

(1) There is a one-to-one mapping of any observable $\mathscr{F}$ on a hermitean operator F of a Hilbert space $\mathscr{H}$.

(2) The observables form a ring on which the commutator can be defined with the following properties:

(a) $[F, G] = FG - GF = -[G, F]$

(b) $[F, \alpha G] = \alpha[F, G]$

(c) $[F, G_1 + G_2] = [F, G_1] + [F, G_2]$

---

[1] See C. Cohen-Tannoudji, B. Diu, F. Laloë, Quantum Mechanics, Wiley Interscience, New York: 1977; R. L. Liboff, Introductory Quantum Mechanics, Holden-Day San Francisco: 1980; G. Ludwig, Foundations of Quantum Mechanics, Springer, New York: 1983.

(d) $(F, G_1G_2) = (F, G_1)G_2 + G_1(F, G_2)$

(e) $(q_j(t), p_k(t)) = \delta_{jk}$

(f) $(p_j(t), p_k(t)) = 0$
(g) $(q_j(t), q_k(t)) = 0$

(3) The time development of an observable is given by
$$\frac{dF}{dt} = (F, H) + \frac{\partial F}{\partial t}$$
where H is the Hamiltonian.

(4) The outcome of a measurement is determined by the expectation value $M(F)$ which satisfies

(a) $M(F^2) \geq 0$
(b) $M(\Sigma \alpha_i F_i) = \Sigma \alpha_i M(F_i)$

[It turns out that
$M(F(q_j, p_j))$
$\quad = \int F(q_j, p_j)\, dm(\mathcal{T})$
where $m(\mathcal{T})$ is a measure of a subspace $\mathcal{T}$ of the phase space $\mathcal{P}$.]

(No equivalent axiom in classical mechanics.)

[Since for any two *subsets*
$\mathcal{R}, \mathcal{S} \subset \mathcal{P}$
$\mathcal{R} = (\mathcal{R} \cap \mathcal{S}) \cup [\mathcal{R} \cap (\mathcal{P} - \mathcal{S})]$
holds true, it follows that any two observables can be measured simultaneously.]

(d) $[F, G_1G_2]$
$\quad = [F, G_1]G_2 + G_1[F, G_2]$
(e) $[Q_j(t), P_k(t)] = i\hbar\delta_{jk}1,$
$\quad \hbar = \frac{h}{2\pi}$

(f) $[P_j(t), P_k(t)] = 0$
(g) $[Q_j(t), Q_k(t)] = 0$

(3) The time development of an observable is given by
$$\frac{dF}{dt} = \frac{1}{i\hbar}[F, H] + \frac{\partial F}{\partial t}$$
where H is the Hamiltonian.

(4) The outcome of a measurement is determined by the expectation value $M(F)$ which satisfies

(a) $M(F^2) \geq 0$
(b) $M(\Sigma \alpha_i F_i) = \Sigma \alpha_i M(F_i)$

[For the special case where the system is described by a function ψ of the Hilbert space of square integrable functions, the expectation value is given by
$M(F) = \int \psi^*(x) F\psi(x)\, dx$.]

(5) Let $\mathcal{R}$ and $\mathcal{S}$ be subspaces of the Hilbert space $\mathcal{H}$. Let $P_{\mathcal{R}}$ be the projection operator on $\mathcal{R}$. Let M_1 and M_2 be the expectation value functions before and after the measurement of the observable $P_{\mathcal{R}}$, respectively. If $M_1(P_{\mathcal{R}}) = 1$, then

(a) $M_2(1 - P_{\mathcal{R}}) = 0$ and
(b) $M_2(P_{\mathcal{S}}) = M_1(P_{\mathcal{S}})$ for any $\mathcal{S} \subseteq \mathcal{R}$.

[Because for any two *subspaces*
$\mathcal{R}, \mathcal{S} \subset \mathcal{H}$
$\mathcal{R} \neq (\mathcal{R} \cap \mathcal{S}) \cup [\mathcal{R} \cap (\mathcal{H} - \mathcal{S})]$
holds true, it follows that $P_{\mathcal{R}}$ and $P_{\mathcal{S}}$ cannot simultaneously be measured in general.]

ANSWERS TO SELECTED PROBLEMS

CHAPTER 2

11. (a) $\dfrac{\pi}{16}$ (b) $\dfrac{\sqrt{2}}{30}$ (c) 2

14. (a) xyz (b) $x \sin y + z \cos y$ (d) $xe^{y} + ye^{-z}$

15. (b) $\begin{pmatrix} 4 & 5 & 6 \\ 7 & 8 & 9 \end{pmatrix}$ (c) $\begin{pmatrix} 3 \\ 4 \end{pmatrix}$ (e) $\begin{pmatrix} 1 & 4 \\ 2 & 5 \\ 3 & 6 \end{pmatrix}$

   (f) $\begin{pmatrix} 17 & 22 & 27 \\ 22 & 29 & 36 \\ 27 & 36 & 45 \end{pmatrix}$ (g) $\begin{pmatrix} 14 & 32 \\ 32 & 77 \end{pmatrix}$ (h) $\begin{pmatrix} 1 & -2 \\ -2 & 4 \end{pmatrix}$

   (i) $\begin{pmatrix} 1 & 2 \\ 2 & 4 \end{pmatrix}$

22. (b) $\lambda_1 = 2 \quad X^{(1)} = \dfrac{1}{\sqrt{13}} \begin{pmatrix} 3 \\ 2 \end{pmatrix}; \quad \lambda_2 = 3 \quad X^{(2)} = \dfrac{1}{\sqrt{2}} \begin{pmatrix} 1 \\ 1 \end{pmatrix}$

   (c) $\lambda_1 = 0 \quad X^{(1)} = \dfrac{1}{\sqrt{6}} \begin{pmatrix} 1 \\ 2 \\ 1 \end{pmatrix}; \quad \lambda_2 = 3 \quad X^{(2)} = \dfrac{1}{\sqrt{3}} \begin{pmatrix} 1 \\ -1 \\ 1 \end{pmatrix};$

     $\lambda_3 = 6 \quad X^{(3)} = \dfrac{1}{\sqrt{2}} \begin{pmatrix} 1 \\ 0 \\ -1 \end{pmatrix}$

CHAPTER 3

1. (a) $\mathbf{v} = (-A\omega \sin \omega t,\ A\omega \cos \omega t,\ B)$,

     $\mathbf{a} = (-A\omega^2 \cos \omega t,\ -A\omega^2 \sin \omega t, 0),\ v = \left[(A\omega)^2 + B^2 \right]^{1/2}$

   (c) $\mathbf{v} = (0,\ A\omega,\ \tfrac{1}{2}A\omega \sin \omega t)$

     $\mathbf{a} = (-A\omega^2(1 + \tfrac{1}{4}\sin^2 \omega t),\ -\tfrac{1}{4}A\omega^2 \sin \omega t \cos \omega t,\ A\omega^2 \cos \omega t)$

     $v = A\omega(1 + \tfrac{1}{4}\sin^2 \omega t)^{1/2}$

5. $t = n\dfrac{\pi}{4}$ even n: maxima, odd n: minima

6. (a) $v = \dfrac{R}{t + \frac{R}{v_0}}$ (b) $a = \dfrac{\sqrt{2}\,R}{\left(t + \frac{R}{v_0} \right)^2}$

12. $v = 2\sqrt{2}\,R\dot{\theta}(1 + \cos \theta)^{1/2}$

CHAPTER 4

3. (a) $\mathbf{a} \times \mathbf{b} = 0$ (b) $b = c = a/2$

4. (a) $-\tfrac{1}{3}(ax^3 + by^3 + cz^3)$ (d) $-k\dfrac{xz}{y^2}$

(e) $-x\sin(yz)$

7. $\mathbf{F} = k\mathbf{r}$

8. $v(t) = v_0 - \dfrac{a}{mb}(e^{bt} - 1), \quad x(t) = \left(\dfrac{a}{mb} + v_0\right)t - \dfrac{a}{mb^2}(e^{bt} - 1)$

$v = 0$ at $t_1 = \dfrac{1}{b}\ln\left(1 + \dfrac{mbv_0}{a}\right)$

$x(t_1) = \dfrac{v_0}{b}\left[-1 + \left(\dfrac{a}{mbv_0} + 1\right)\ln\left(\dfrac{mbv_0}{a} + 1\right)\right]$

9. $v(t) = v(0) + \dfrac{F_0}{m}\dfrac{1}{(\lambda^2 + \omega^2)}[-(\lambda\sin(\omega t + \delta) + \omega\cos(\omega t + \delta))e^{-\lambda t}$
$\qquad\qquad\qquad\qquad\qquad\qquad\qquad + \lambda\sin\delta + \omega\cos\delta]$

$v(\infty) = v(0) + \dfrac{F_0}{m}\dfrac{1}{(\lambda^2 + \omega^2)}(\lambda\sin\delta + \omega\cos\delta)$

10. $v(\infty) = \dfrac{mg}{b}$

11. (c) $H = \dfrac{m}{2b}\ln\left(1 + \left(\dfrac{v_0}{\beta}\right)^2\right), \quad \beta^2 = \dfrac{mg}{b}$

14. (c) $F(t) = -nmb^2[x_0^{n+1} + (n + 1)bt]^{-(2n+1)/(n+1)}$

16. (a) $x = x_0\left[1 + \left(1 + \dfrac{a}{m}\right)\dfrac{v_0}{x_0}t\right]^{m/(a+m)}$

(b) set $k^2 = \dfrac{2mv_0}{b} - x_0^2$

if $k^2 > 0$: $x = k\dfrac{x_0 + k\tan\frac{kb}{2m}t}{k - x_0\tan\frac{kb}{2m}t}$

if $k^2 = 0$: $x = \dfrac{x_0}{\frac{x_0 b}{2m}t + 1}$

if $k^2 < 0$: $x = \sqrt{-k^2}\,\dfrac{1 + e^{\sqrt{-k^2}\frac{b}{m}t}}{1 - e^{\sqrt{-k^2}\frac{b}{m}t}}$

CHAPTER 5

4. $A_2 = \dfrac{\sqrt{13}}{2}A_1$

5. (b) $k = 12B\left(\dfrac{A}{3B}\right)^{1/4}$ (d) $k = 2\dfrac{A}{a^2}$ (f) $k = b^2\sqrt{A^2 - B^2}$

8. $\bar{T} = \bar{V} = \frac{1}{2}E$

12. (b) $A_+ = A_-^* = \dfrac{1}{2}\left(-1 + i\dfrac{\gamma}{\omega}\right)A$

13. $x(t) = \dfrac{F_0}{m\omega_0^2}(\sin\omega\tau\sin\omega t + (\cos\omega\tau - 1)\cos\omega t), \quad t > \tau$

CHAPTER 6

3. (a) $f(r) = \dfrac{l^2}{mA^{2n}}(n - 1)r^{2n-3}$

(b) $f(r) = -\dfrac{2l^2}{m}A^2 r^{-5}$

6. If $b^2 \equiv 1 - \dfrac{mk}{l^2} > 0$: $r = (A\cos b\theta + B\sin b\theta)^{-1}$

9. (a) $\varepsilon = \dfrac{v_1}{v_0}$

11. $T = 2\pi\sqrt{\dfrac{GM}{R^3}}$

14. (c) $\sigma(\theta) \approx \dfrac{\pi}{8}\dfrac{k}{E}\dfrac{1}{\theta^3}$

15. (d) $\sigma_{\text{total}} = 2\pi a^2 \dfrac{E + V_0}{2E + V_0}$

CHAPTER 7

8. (a) $T = \sqrt{\dfrac{2h}{g}}\left(\dfrac{1 + \varepsilon}{1 - \varepsilon}\right)$ (b) $H = h\left(\dfrac{1 + \varepsilon^2}{1 - \varepsilon^2}\right)$

12. (a) $Q = m\bar{v}_1\bar{v}_2\cos\psi$

15. $\varphi = \tfrac{1}{2}\varphi'$

16. (a) $\theta = \varphi = 30°$ (b) $\bar{T}_d = 1\text{ MeV}$, $\bar{T}_p = 2\text{ MeV}$

CHAPTER 8

4. $x_1 = \dfrac{m_1^2}{m_1^2 + m_2^2}(l - 2a)$, $x_2 = \dfrac{m_2^2}{m_1^2 + m_2^2}(l - 2a)$

5. $T = \tfrac{1}{2}ma^2(\sinh^2 u + \sin^2 v)(\dot{u}^2 + \dot{v}^2)$

9. $\ddot{y} - \dfrac{g}{l}y = 0$

10. $\Lambda_1 L = (m_1 l_1^2 + m_2 l_2^2)\ddot{\theta}_1 + m_2 l_1 l_2\cos(\theta_2 - \theta_1)\ddot{\theta}_2$
$\qquad - m_2 l_1 l_2\sin(\theta_2 - \theta_1)\dot{\theta}_2^2 + (m_1 l_1 + m_2 l_2)g\sin\theta_1 = 0$
$\quad \Lambda_2 L = m_2 l_1 l_2\cos(\theta_2 - \theta_1)\ddot{\theta}_1 + m_2 l_2^2\ddot{\theta}_2$
$\qquad + m_2 l_1 l_2\sin(\theta_2 - \theta_1)\dot{\theta}_1^2 + m_2 l_2 g\sin\theta_2 = 0$

13. (e) $(m_1 + m_2)\ddot{r} - m_1 r\dot{\theta}^2 - m_1 g\cos\theta = 0$
$\qquad r\ddot{\theta} + 2\dot{r}\dot{\theta} + g\sin\theta = 0$

CHAPTER 9

5. (a) $I'_{12}I'_{23}I'_{31} > 0$, $I_{11} = I'_{11} + \dfrac{I'_{13}I'_{23}}{I'_{12}} + \dfrac{I'_{12}I'_{23}}{I'_{13}}$
$\quad$ (b) $\mathbf{R} = (1/\sqrt{3})(1, 1, 1)$, $I_{kl} = 7\delta_{kl}$

9. $I = \tfrac{3}{5}MR^2$

12. $y = \dfrac{1}{12}\dfrac{3l^2 + x^2}{l}$, $m = \dfrac{24lx}{3l^2 + x^2}M$

14. $f = \dfrac{1}{2\pi} \sqrt{\dfrac{g}{2k_{CM}}}$

22. $\omega^2 = \dfrac{2T(I_1 + I_3) - L^2}{I_1 I_3}$

26. The angle $\sphericalangle(\mathbf{e}_1, \mathbf{e}_1') = 82°$ is not the angle γ of the combined finite rotation, $\gamma = 98°$.

CHAPTER 10

1. (a) $\omega^2 = \frac{1}{2}(T_{11}T_{22} - T_{12}^2)^{-1}\{(V_{11}T_{22} - 2V_{12}T_{12} + V_{22}T_{11})$
$\pm [(V_{11}T_{22} - V_{22}T_{11})^2 + 4(V_{11}T_{12} - T_{11}V_{12})(V_{22}T_{12} - T_{22}V_{12})]^{1/2}\}$

$\quad$ **(b)** $\omega^2 = \dfrac{1}{2}\left\{\dfrac{V_{11}}{T_{11}} + \dfrac{V_{22}}{T_{22}} \pm \left[\left(\dfrac{V_{11}}{T_{11}} - \dfrac{V_{22}}{T_{22}}\right)^2 + 4\dfrac{V_{12}^2}{T_{11}T_{22}}\right]^{1/2}\right\}$

2. (a) $x_1 = x_3 = \sqrt{\dfrac{2k_2}{k_1}}\, x_2$

$\quad$ **(b)** $\omega_1 = 0 \qquad \omega_2 = \sqrt{\dfrac{k_1}{m}} \qquad \omega_3 = \sqrt{\dfrac{k_1 + k_2}{m}}$

5. $\omega_1 = \sqrt{\dfrac{k_1}{m}} \qquad \omega_2 = \sqrt{\dfrac{k_1 + 2k_2}{m}}$
$\qquad a_{11} = a_{21} = a_{12} = -a_{22} = (2m)^{-1/2}$

6. $\omega_1 = 0 \qquad \omega_2 = \sqrt{\dfrac{k}{m}} \qquad \omega_3 = \sqrt{\dfrac{k}{m}\left(1 + \dfrac{2m}{M}\right)}$

$\mathbf{X}^{(1)} = \dfrac{1}{\sqrt{2m + M}}\begin{pmatrix} 1 \\ 1 \\ 1 \end{pmatrix}$ (translation)

$\mathbf{X}^{(2)} = \dfrac{1}{\sqrt{2m}}\begin{pmatrix} 1 \\ 0 \\ -1 \end{pmatrix} \qquad \begin{array}{ccc} \leftarrow & \cdot & \rightarrow \\ 1 & 2 & 3 \end{array}$

$\mathbf{X}^{(3)} = \dfrac{1}{\sqrt{2m}}\dfrac{1}{\sqrt{1 + 2m/M}}\begin{pmatrix} 1 \\ -2m/M \\ 1 \end{pmatrix} \qquad \begin{array}{ccc} \rightarrow & \leftarrow & \rightarrow \\ 1 & 2 & 3 \end{array}$

7. $\omega_1 = \sqrt{\dfrac{k_1}{m}} \qquad \omega_2 = \sqrt{\dfrac{k_1 + k_2}{m}}$
$\qquad a_{11} = a_{21} = a_{12} = -a_{21} = (2m)^{-1/2}$

index

Pages in *italics* refer to sections or whole paragraphs where the quantities are discussed more thoroughly.